AN INTRODUCTION TO SINGLE MOLECULE BIOPHYSICS

Foundations of Biochemistry and Biophysics

Currently available

Quantitative Understanding of Biosystems: An Introduction to Biophysics
Thomas M. Nordlund

Biomolecular Thermodynamics: From Theory to Application
Douglas E. Barrick

Biomolecular Kinetics: A Step-by-Step Guide
Clive R. Bagshaw

An Introduction to Single Molecule Biophysics
Yuri L. Lyubchenko (Ed.)

Forthcoming

Physical Principles in Nucleic Acid Chemistry
David E. Draper

RNA Biophysics
Kathleen B. Hall

AN INTRODUCTION TO SINGLE MOLECULE BIOPHYSICS

Edited by
Yuri L. Lyubchenko
College of Pharmacy
University of Nebraska Medical Center

CRC Press
Taylor & Francis Group
Boca Raton London New York

CRC Press is an imprint of the
Taylor & Francis Group, an **informa** business

Cover: Image by Salvatore Chiantia, courtesy of Petra Schwille, Max Planck Institute of Biochemistry.

CRC Press
Taylor & Francis Group
6000 Broken Sound Parkway NW, Suite 300
Boca Raton, FL 33487-2742

First issued in paperback 2018

ISBN 13: 978-1-138-61596-0 (pbk)
ISBN 13: 978-1-4398-0694-4 (hbk)

Library of Congress Cataloging-in-Publication Data

Names: Lyubchenko, Yuri L., editor.
Title: Introduction to single molecule biophysics / [edited by] Yuri L. Lyubchenko.
Other titles: Foundations of biochemistry and biophysics.
Description: Boca Raton : Taylor & Francis, 2017. | Series: Foundations of biochemistry and biophysics
Identifiers: LCCN 2017033078 | ISBN 9781439806944 (hardback : alk. paper)
Subjects: | MESH: Single Molecule Imaging--methods | Biophysical Phenomena
Classification: LCC QH345 | NLM WN 190 | DDC 572--dc23
LC record available at https://lccn.loc.gov/2017033078

Visit the Taylor & Francis Web site at
http://www.taylorandfrancis.com

and the CRC Press Web site at
http://www.crcpress.com

Contents

Series Preface

Biophysics encompasses the application of the principles, tools, and techniques of the physical sciences to problems in biology, including determination and analysis of structures, energetics, dynamics, and interactions of biological molecules. Biochemistry addresses the mechanisms underlying the complex reactions driving life, from enzyme catalysis and regulation to the structure and function of molecules. Research in these two areas is having a huge impact in pharmaceutical sciences and medicine.

These two highly interconnected fields are the focus of this book series. It covers both the use of traditional tools from physical chemistry, such as nuclear magnetic resonance (NMR), X-ray crystallography, and neutron diffraction, as well as novel techniques, including scanning probe microscopy, laser tweezers, ultrafast laser spectroscopy, and computational approaches. A major goal of this series is to facilitate interdisciplinary research by training biologists and biochemists in quantitative aspects of modern biomedical research, and teaching core biological principles to students in physical sciences and engineering.

Proposals for new volumes in the series may be directed to Lu Han, Executive editor, at CRC Press/Taylor & Francis Group (lu.han@taylorandfrancis.com).

Preface

Single molecule biophysics (SMB) is a relatively new branch of molecular biophysics, which utilizes a set of methods that enable imaging or probing of individual molecules. The attractiveness of SMB is defined by two major features. The first is biological systems, including cells, which consist of a limited number of biological molecules or complexes. A study of this kind requires techniques capable of probing single molecules. Second are the intricate functions of biological molecules and systems as they assemble in transient states. Characterization of these states is critical in understanding the molecular mechanism of the biological process under study. As they become available, SMB techniques have been applied to numerous biological problems associated with understanding the molecular mechanisms of DNA replication, transcription, and translation, as well as the functions of molecular machines. Many of these examples and others are discussed in the book in association with a specific method.

The discoveries of major SMB techniques have historically been associated with Nobel Prizes. For example, the atomic force microscope (AFM) belongs to the family of scanning probe microscopes, the discovery of which was awarded the Nobel Prize in Physics in 1986 to G. Binnig and H. Rohrer at IBM, Zurich, who published their discovery only 4 years prior to the award. Current single molecule fluorescence methods originate from the 1989 publication by W. Moerner and L. Kador, with a Nobel Prize awarded to W. Moerner in 2014 for the discovery. Optical traps or tweezers utilize the phenomenon of manipulating atoms and small particles by focused light beam. By applying this method in physics to the cooling and trapping of atoms with laser light, the assembly of a Bose–Einstein condensate was achieved, leading to Nobel Prizes in Physics in 1997 and 2001.

This introductory book broadly covers the motivations, history, and methodologies of the major single molecule imaging and probing techniques, which can be applied to numerous important biological systems and questions.

Furthermore, this book highlights the most useful and exciting strengths of these evolving methods through examples of several recent biological applications. The primary target audience for this book is undergraduate and graduate students, academics, and professionals who are looking to learn more about this rapidly growing field and may benefit from some of its many research capabilities. An important focus of the book is to introduce the key concepts and ideas of the major approaches in the single molecule field in a relatively simple manner. A defining characteristic of the book's style is that large sets of specific examples have been assembled for each biological application described. The goal of this layout is to better familiarize readers with techniques they may be interested in pursuing for their own research. Additionally, up-to-date information about several different applications is discussed, and a substantial reference list is provided, which even seasoned practitioners in the field should find useful both as a personal reference and as an introductory guide to the field for incoming students and colleagues.

The book is assembled around six different SMB techniques, with each chapter dedicated to a specific method. The first chapter covering single molecule fluorescence techniques is written by Deniz. This chapter provides basic knowledge in fluorescence and the Förster resonance energy transfer (FRET) phenomenon, and describes principles of instrumentation enabling implementation of the techniques at the single molecule level. Brief discussions of applications of the methods round out the chapter and highlight the power of the methods to answer fundamental questions in biophysics and biology. In Chapter 2, Lyubchenko provides an overview of AFM as a tool for single molecule imaging. In addition to a description of the instrument, its main properties, and the spatial resolution limit, the chapter describes the use of AFM for direct time-lapse visualization of dynamics of molecular complexes at the nanoscale. A great deal of attention was given to high-speed AFM capable of data acquisition at a video rate. This emerging AFM technique has already produced new knowledge in a large number of applications, with a few of them described. Josephs, Marszalek, and Scholl in Chapter 3 review the applications of AFM as a nanotool for measuring the interactions between molecules and the mechanical properties of isolated molecules and their complexes. The authors describe the principles of AFM single molecule force spectroscopy (SMFS) experiments, interpretation of data, and general principles of force spectroscopy data analysis. Although AFM force spectroscopy is a mode of operation of essentially all commercially available AFM instruments, the authors describe principles for designing a force spectroscopy instrument. As in other chapters, a few applications of this instrumentation are described. In Chapter 4, Bianco, Lyubchenko, and Sun describe the principles and operation of magnetic tweezers. In this technique, the system under study to which a small magnetic bead is attached is interrogated with an external magnet. This method, essentially established and developed by Bensimon and Croquette in France, caught the broad attention of the SMB community due to the relative simplicity of the instrument construction. Compared with other probing techniques, AFM and optical tweezers, the major attractive feature of magnetic tweezers is the ability to control the rotational dynamics of the system. Advances in this application of magnetic tweezers are described, and a few applications are presented. Capitanio outlines the principles of optical tweezers methodology in Chapter 5. This is one of the most widely used single molecule techniques that attracts investigators through its ability to probe numerous biological systems, discover their dynamic behavior, and investigate force dependence. Principles of the instrument assembly, the fundamentals of force and position measurements, and the operation of optical tweezers in different configurations are described. Capitanio guides the researcher to the most appropriate optical tweezers configuration to be used with a given biological system, how to optimize the measurements, and what resolution limits should be expected. Most examples are presented on the application of optical tweezers to single biological molecules *in vitro*, with a final illustration of recent applications of the technique to quantitative studies to probing a single cell. Protein nanopores occupy a central role in biology as the primary conduit for a cell's communication with the outside world, and synthetic nanopores have long aimed to replicate this functionality in engineered assemblies. In Chapter 6, Noy and Wanunu go over the basic concepts of nanopore transport, and the most common types of biological and artificial nanopores, including the recently developed carbon nanotube porins. They also review some of the unique transport phenomena that are observed in these structures, and describe the physical and technological basis of nanopore-based sensing. Particular emphasis is given to the nanopore-based DNA sequencing technology that will be underpinning the next generation of genomics and health care applications.

The set of techniques assembled in this book includes the most widely used experimental SMB experiments. These methods have already led to a large number of discoveries, with some selected examples described in this book. A major trend in the SMB field is to utilize a combination of the described techniques in one integrated instrument, and a few such integrations are described in this book. However, even these methods are in flux and undergo constant improvements and modifications. Novel methods appear as well, and these advances set the basis for future editions of such an introductory book.

Yuri L. Lyubchenko
University of Nebraska Medical Center

Editor

Yuri L. Lyubchenko is professor of Pharmaceutical Sciences at the University of Nebraska Medical Center, Omaha, Nebraska. His research focuses on understanding fundamental mechanisms underlying health and disease, which are key to developing new and more effective diagnostics and medications. This primarily basic research allows him not only to identify new drug targets for small molecule drugs, but also to develop the nanotools and methods needed to discover novel approaches for diagnosis, treatment, and disease prevention, and to more rapidly determine their efficacy at the molecular level.

Contributors

Piero R. Bianco
Department of Microbiology and Immunology
Center for Single Molecule Biophysics
University at Buffalo
Buffalo, New York

Marco Capitanio
LENS (European Laboratory for Non-Linear Spectroscopy)
University of Florence
Florence, Italy

Ashok Deniz
The Scripps Research Institute
La Jolla, California

Eric A. Josephs
Department of Mechanical Engineering and Materials Science
Duke University
Durham, North Carolina

Yuri L. Lyubchenko
Department of Pharmaceutical Sciences
College of Pharmacy
University of Nebraska Medical Center
Omaha, Nebraska

Piotr E. Marszalek
Department of Mechanical Engineering and Materials Science
Duke University
Durham, North Carolina

Aleksandr Noy
Lawrence Livermore National Laboratory
Livermore, California

and

University of California, Merced
Merced, California

Zackary N. Scholl
Department of Computational Biology and Bioinformatics
Duke University
Durham, North Carolina

Zhiqiang Sun
Department of Pharmaceutical Sciences
College of Pharmacy
University of Nebraska Medical Center
Omaha, Nebraska

Meni Wanunu
Northeastern University
Boston, Massachusetts

1 Single Molecule Fluorescence

Ashok Deniz
The Scripps Research Institute
La Jolla, California

Contents

The past couple of decades have seen a veritable explosion of single molecule studies of biological phenomena, along with concurrent improvements in the technologies needed for such studies. Such studies provide several substantial benefits compared with standard ensemble studies, and on a practical level they provide complementary information for answering important biophysical and biological questions. For example, these methods have provided important new insights into the functioning of enzymes, dynamic structural complexity in proteins and nucleic acids, protein folding and misfolding, the function of motor and gene regulation proteins, and even the entry and dynamics of viruses into cells.

Single molecule methods offer several important strengths. While many ensemble methods average properties over millions or billions of molecules resulting in information loss, Single molecule methods measure the properties of molecules one at a time, allowing more detailed information to be recorded. For example, conformational distributions of proteins and nucleic acids can be recorded. Moreover, these molecules are often quite dynamic, and single molecule methods can better capture heterogeneity in these dynamics. In this regard, most importantly, complex, multistate, stochastic dynamics can be captured, which would be virtually impossible in ensemble experiments because different molecules would be proceeding through different steps of reaction trajectories, even if all reactions were initiated at the same time. In this chapter, we discuss some of the important single molecule fluorescence methods and follow up with short discussions of some interesting biological problems that have been tackled using these methods.

Figure 1.1 illustrates a few important parameters that are of interest to students and practitioners of biophysics in its broadest context (including biologists, physicists, and chemists). Single molecule methods can provide information about many such biological properties over a large dynamic range, mainly *in vitro*, but increasingly in living cells and *in vivo* (Deniz et al. 2008). For example, distance measurements can be performed with resolutions from mesoscopic to atomic dimensions by using different methods. The single molecule practitioner will ideally choose a particular method to be used based on the specific problem or biological system under study. Here, we first briefly list several molecular properties of interest, along with a brief discussion of a few common single molecule methods that can be used to make measurements of these properties.

In the next section, we introduce individual types of common single molecule measurements briefly, highlighting their importance in biophysical and biological studies.

1.1 Single Molecule Fluorescence Observables and Detection Modes

1.1.1 Observables

1.1.1.1 Fluorescence

Fluorescence is a relaxation process available to single electronic excited states of molecules, and this process involves the emission of a photon with an energy corresponding to the energy difference between the excited state and the ground state (Figure 1.2). Typically, emission occurs from a lower vibrational level of the excited state; hence, the emission is red shifted in wavelength

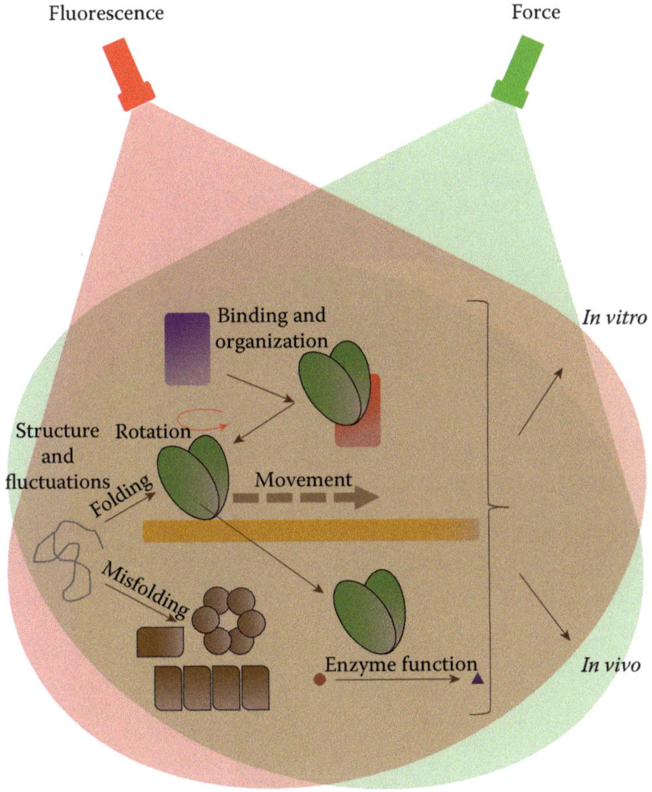

Figure 1.1 Single molecule fluorescence (and force/manipulation) methods provide a wide variety of information about biological systems. In this chapter, we focus on fluorescence methods.

compared to the excitation light. Because of this feature, the excitation light can be blocked while letting the emission light be detected, which is a huge advantage, contributing to the very high sensitivity of the method for single molecule measurements.

In addition, the timescale of fluorescence is such that it typically competes with other excited state relaxation processes, which in turn allows it to report molecular properties of interest. For example, relaxation can also occur by non-radiative means; hence, the "brightness" of the fluorescence is often modulated by intramolecular or intermolecular quenchers, thus reporting on the proximity of such quenchers. In the special case of Förster resonance energy transfer (FRET), the proximity of a second chromophore allows longer range distances to be measured. Tumbling of large molecules can also occur on a timescale similar to fluorescence, which allows for the measurement of such tumbling and molecular size via polarization measurements (Figure 1.3). Finally, photon emission can be used to localize the location of molecules down to nanometer (10^{-9} m)

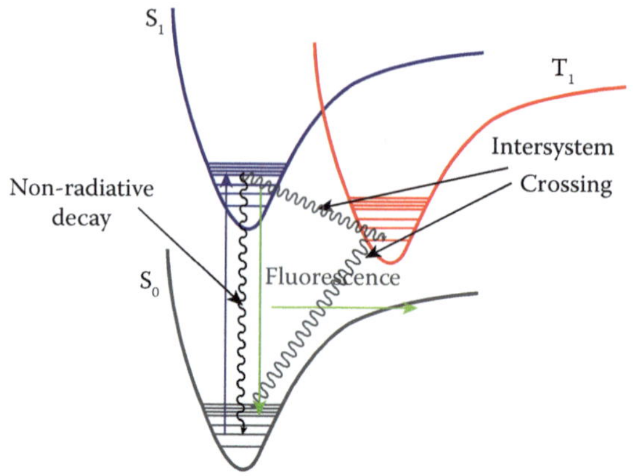

Figure 1.2 Jablonski diagram depicting fundamental electronic states and processes relevant to fluorescence.

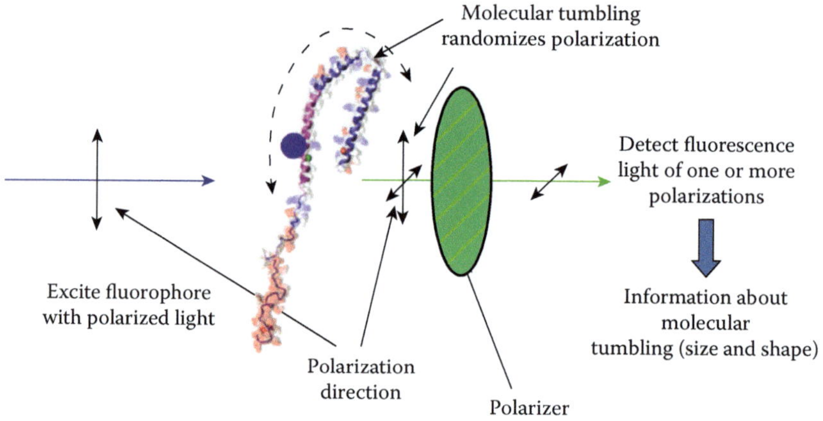

Figure 1.3 The principle of fluorescence polarization experiments and measurement of molecular tumbling.

precision, which can be used to follow the details of movement of molecules *in vitro* and in live cells.

1.1.1.2 Single-Dye Measurements (Intensity and Other Properties)

A number of single molecule measurements have used fluorescence emission from a single dye to report on molecular properties. For example, in pioneering single molecule experiments on enzyme molecules, Xie and coworkers used the fact that a redox enzyme cofactor's fluorescence was turned on and off depending on whether it was oxidized or reduced (Lu et al. 1998). Since the enzymatic cycle involved oxidation and reduction of the cofactor, recording the "blinking" in emission intensity of the cofactor fluorescence allowed them to record the catalytic activity of these enzymes in real time. Even before these measurements, others had used a substrate that turned fluorescent upon reaction to the product to measure the activity of individual enzymes (Xue and Yeung 1995; Craig et al. 1996), although here individual turnovers were not recorded. These kinds of experiments have resulted in the generation of entirely new kinds of information about the complexity of enzyme function.

Emission from one kind of dye (but from multiple molecules) has also been used in pioneering experiments, for example, to observe the rotation of molecular motors. Finally, photons have a direction associated with them (polarization). This feature can be used to measure the orientation or rotation of molecules, which has also been used in a number of single molecule experiments. For example, polarization information can report on the size of a molecule, because tumbling rate scales with the size, and tumbling of the molecule following excitation and before emission results in randomization of the polarization of the emitted light. Molecular size can give information about conformation and binding—indeed, ensemble polarization is commonly used as a sensitive tool for screening binding, for example, in the search for drug candidates.

Yet another useful single-dye analysis that has been used extensively is fluorescence correlation spectroscopy (Magde et al. 1972; Rigler et al. 1993; Eigen and Rigler 1994; Thompson et al. 2002; Kim et al. 2007). In some ways, this was a precursor of modern fluorescence single molecule experiments. The experiment essentially consists of using confocal detection to monitor emission from a small number of molecules as they diffuse into and out of a focal volume. A correlation analysis then is used to extract information about the fluctuation timescales and amplitudes from the apparently noisy data (Figure 1.4). This method has been widely used to study diffusion times (and hence binding or sometimes conformational changes). It can also be used to study chain fluctuations using an appropriate quencher, as for the disordered state of the protein Sup35-NM (Figure 1.4). More generally, this represents a class of correlation or fluctuation analysis methods.

1.1.1.3 Single Molecule FRET (Two or More Dyes)

Single molecule FRET is the single molecule analog of the well-known ensemble technique commonly referred to as FRET (Förster Resonance Energy Transfer) that is useful for measuring distances (and, hence, structure) in molecules. Named after Theodor Förster, it is also commonly known as fluorescence resonance energy transfer, which may be a bit misleading because transfer is nonradiative and not via the emission and reabsorption of a photon. Pioneering work several decades ago by Stryer and Haughland resulted in demonstration of the strong distance dependence of the transfer (Stryer and Haugland 1967). In the mid-1990s to late 1990s, the first demonstration of single molecule FRET (Ha et al. 1996) and its distance dependence (Deniz et al. 1999) were followed by quite extensive use of the method in the single molecule community.

As shown in Figure 1.5, the phenomenon refers to the nonradiative transfer of excitation energy from a donor molecule to an acceptor molecule by the coupling

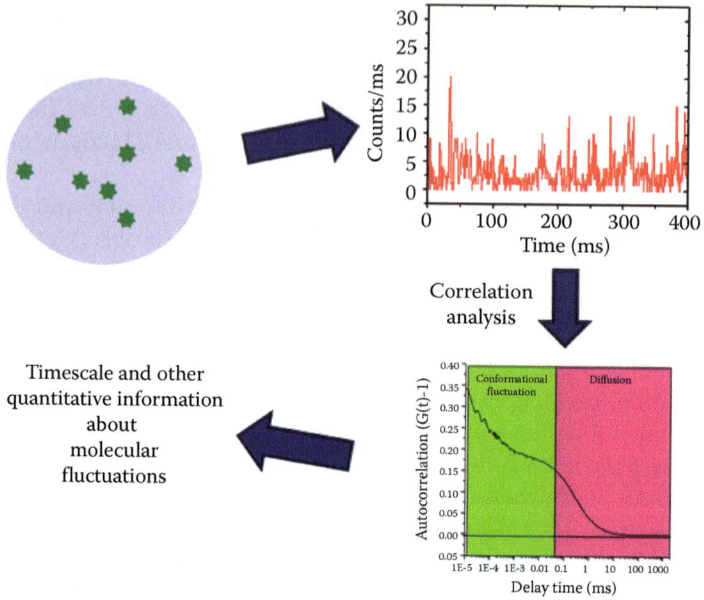

Figure 1.4 The basic principles behind fluorescence correlation spectroscopy measurements. (Adapted from Mukhopadhyay, S., et al., *Proc. Natl. Acad. Sci. U S A*, 104, 2649, 2007.)

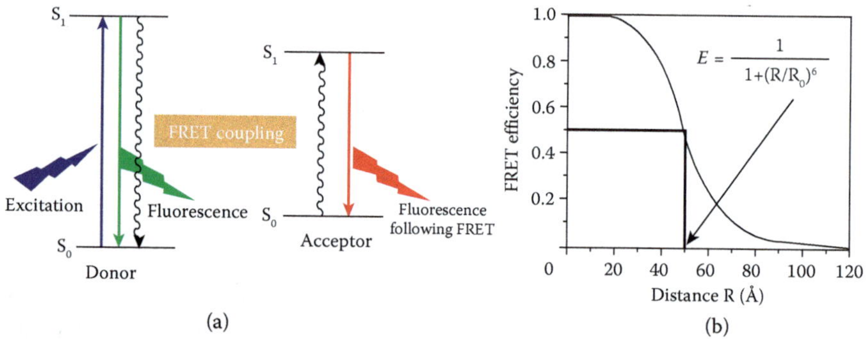

Figure 1.5 Jablonski diagram of the FRET process (a) and distance dependence of FRET (b).

between their transition dipoles. The transfer rate has a strong distance dependence (r^6). This translates into a strong distance dependence for the transfer efficiency (E_{FRET}), which is often measured in experiments. E_{FRET} represents the fraction of donor excited states that result in energy transfer to the acceptor (vs. fluorescence of the donor or nonradiative decay). As can be seen, the transfer efficiency, which is a function of a competition in rate constants for the various decay processes of the donor, can report on the distance between donor and acceptor dyes (30–70 Å range).

By labeling a protein, RNA, or other molecule with such donor and acceptor dyes, it is then possible to measure distance and hence conformational properties for these molecules at single molecule resolution. Labeling methods and instrumentation are briefly discussed in Sections 2.2.3 and 2.2.2, respectively.

As an example, we briefly discuss the simple case of the first direct observation of two-state protein folding by FRET by Deniz et al. (2000). The protein studied was chymotrypsin inhibitor 2 (CI-2), which was a protein believed to fold via a two-state folding mechanism. In other words, the protein resided either in a denatured state or a folded state, but not significantly in intermediate conformations. To carry out this study, the protein was first produced in a dual-labeled form with donor and acceptor sites at positions expected to report on denatured and folded states with different FRET signatures. Site-specific labeling was achieved using synthetic and ligation methods (Deniz et al. 2000). Freely diffusing proteins were examined; hence, potential perturbation due to interactions with surfaces was avoided. Each single molecule experiment consisted of measuring FRET efficiencies for a large number (>1000) of single molecules and collecting these as a histogram depicting the number of events versus E_{FRET}. As the examples in Figure 1.6 show, these histograms provide information about populations of proteins in solution.

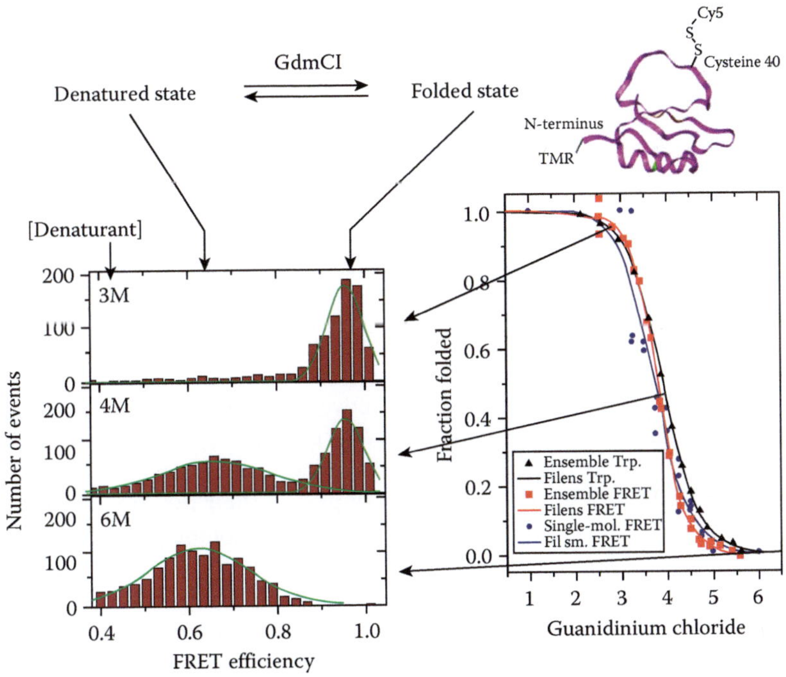

Figure 1.6 Two-state protein folding directly validated by smFRET experiments. (Adapted from Deniz, A. A., et al., *Proc. Natl. Acad. Sci. U S A*, 97, 5179, 2000.) Shown are the two-state model; protein with attached dyes; smFRET histograms at low, intermediate, and high denaturant concentration; and extracted denaturation curves with ensemble curves for comparison.

Under native conditions (absence of denaturant), proteins adopted a high E_{FRET} or compact conformation as expected. Under unfolding conditions, a peak was observed ~0.65 E_{FRET}, which corresponded to a more expanded denatured state. Interestingly, under intermediate levels of denaturant, the experiment provided snapshots of either folded or denatured proteins, with no significant intermediates observed, thereby providing a direct observation of a two-state folding mechanism. An interesting feature was that a denaturation curve could be directly extracted from the data by using areas of the peaks (as a control, curve corresponded well with a denaturation curve from ensemble data on an unlabeled protein calculated using a two-state model). In addition, there may have been a small shift in the denatured peak position, which would indicate that the denatured state was expanding as a function of increasing denaturant, although the small shift did not allow this conclusion to be made unequivocally. More recent experiments on other protein systems have revealed more dramatic shifts, for example, in the case of an intrinsically disordered yeast prion protein monomer (Mukhopadhyay et al. 2007). Furthermore, it should be noted that the denatured and native states can be more generally described as ensembles in structures rapidly interconverting on the timescale of these experiments. Experiments directly probing the fluctuations within these ensembles have been carried out using correlation techniques.

This single molecule FRET (smFRET) method has found wide use in the single molecule community, with applications ranging from probing folding of proteins and complexes, movement of molecular motors on their DNA or other tracks, and formation or dissociation of complexes *in vitro* and in cells. More recently, this method has even been extended to additional dimensions, for example, three-color smFRET (Hohng et al. 2004; Clamme and Deniz 2005; Gambin and Deniz 2010). These methods will allow more global analysis of molecules and complexes at single molecule resolution.

1.1.1.4 Single Molecule Localization and Tracking

Another important feature of single molecule experiments is the possibility of tracking the position of a molecule as it moves (Brandenburg and Zhuang 2007; Greenleaf et al. 2007; Chang et al. 2008). For example, the movement of protein motors on molecular tracks, of protein receptors on cell membranes, or cellular internalization of viruses can be a key part of their functions or infection; hence, a huge effort has gone into understanding the complicated characteristics of these motions. Tracking in its simplest form consists of monitoring photons from a single molecule using a high-sensitivity charge-coupled device (CCD) camera. Since molecules are typically much smaller than the wavelength of light (~450–650 nm for visible light used in many single molecule experiments), the detected emission profile (point spread function) is determined by the optical properties of the excitation and collection. This can often be approximated reasonably by a two-dimensional (2D) Gaussian (Figure 1.7), which is used to fit and pinpoint the location of the molecule. By performing this procedure as a function of time, the path of motion of the molecule can be determined with nanometer precision.

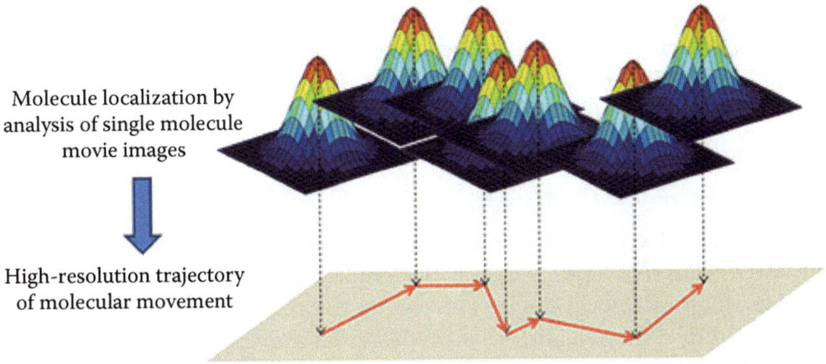

Figure 1.7 Single-particle or single molecule tracking. At the top are 2D Gaussian fits of image data; the center of the 2D Gaussian provides a highly accurate measure of the particle position, as depicted in the trajectory in red.

A related set of methods that has emerged over the last few years are the so-called super-resolution imaging methods. This is a neat way to get around the classical diffraction limit of imaging, which was the dogma in the imaging field until about a decade back. The idea is that diffraction limits the resolution of two objects below about half the wavelength of the emitted light. However, as noted above, if one images the objects (molecules) one at a time, their positions can be localized with much higher precision (arbitrarily high, depends on the number of collected photons). Hence, groups (initially of Xioawei Zhuang [Rust et al. 2006] and Eric Betzig [Betzig et al. 2006]) have developed methods to stochastically turn on a small fraction of single molecules in a sample, localize them by imaging, turn them off, then turn on another small subset, and so on. Hence, by imaging an ensemble of molecules, one at a time, high-resolution imaging can be performed, which is providing detailed views of cellular architecture that were previously unavailable from fluorescence imaging (Bates et al. 2008; Patterson et al. 2010). Indeed, the impact of this advance is large enough that the 2014 Nobel Prize in Chemistry in part went to Betzig for his above contribution.

1.1.2 Instrumentation and Measurement Modalities

Several types of instruments are used for single molecule fluorescence measurements. These were often lab assembled until recently, although some are now available commercially. Key features required for single molecule measurements are high-quality illumination and detection to allow rejection of background photons, and high-efficiency detection of the small number of photons emitted by single molecules. The light source for excitation is usually a laser, which provides illumination with advantageous optical properties, including beam profile and narrow spectral bandwidth. The former permits focusing the beam to a very small volume, and the latter permits better rejection of the intense excitation light prior to measurement of the very low-intensity single molecule fluorescence light. The excitation geometry usually seeks to limit the detection area or volume to nanometer-micron dimensions. Correspondingly, measurement modes include total internal reflection fluorescence (TIRF), confocal and near-field detection, and epi imaging. Detection is via high-efficiency detectors such as high-end CCD cameras (TIRF and epi) and avalanche photodiodes (APDs, confocal). Confocal and TIRF microscopies are the most commonly used modes, and these are discussed in more detail later.

1.1.2.1 Total Internal Reflection Fluorescence

Total internal reflection (TIRF) is a phenomenon where light passing from a higher to a lower refractive index medium is completely reflected if the incident angle is greater than a "critical angle" defined by the difference in refractive indices. Although 100% reflection is achieved in the far field, a very narrow evanescent electromagnetic field is generated at the far side of the interface. This field decays exponentially with a length scale of ~100 nm. The basic features of a TIRF instrument are depicted in Figure 1.8. Laser light passes through a coupling prism, coupling medium, and fused silica slide, then encounters an interface with a lower refractive index medium (aqueous sample). If the incident angle is adjusted appropriately, TIR occurs, generating a thin layer of illumination in the sample solution. The molecules of interest are usually immobilized on the slide and are illuminated by the evanescent field. Fluorescent impurities and so on in the solution that are outside of the evanescent field are not illuminated and hence do not contribute to fluorescent background. Fluorescence emitted by the single molecules is then collected by a high numerical-aperture objective, which permits high collection efficiency. The light then passes through filters to reject excitation light; additional optics (e.g., to separate different wavelength regions for FRET detection) is then focused onto a CCD camera to produce an image. Intensified CCD cameras were often used in the past; these have typically been replaced by electron-multiplying CCDs (with better

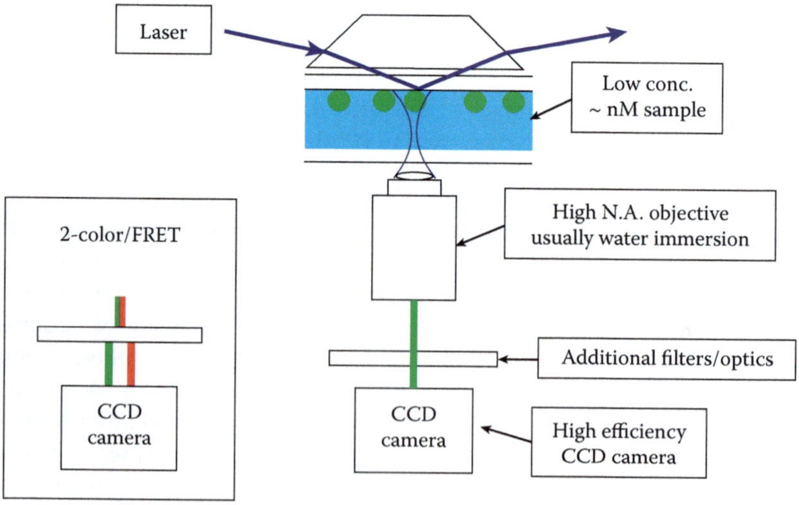

Figure 1.8 The basic instrumental setup for total internal reflection fluorescence (TIRF) single molecule experiments.

detection characteristics), and newer cameras and detectors are under development. The camera provides an image of the detection area, with bright spots corresponding to signals from single molecules. A movie of such images then provides temporal information, typically with a time resolution of 10–100 ms. Analyses of the movies provide time trajectories about the fluorescence characteristics and hence molecular properties of these single molecules. For example, a FRET experiment can provide information about distance as a function of time, and provide information about conformational fluctuations of the molecules of interest. An important characteristic of these types of measurements is the ability to follow hundreds of single molecules simultaneously and in real time for extended periods (usually up to seconds or minutes). Detailed analyses can be performed on the detected time trajectories. Common analyses for smFRET experiments include determining donor and acceptor intensity trajectories, converting them to a FRET time trajectory, assigning states to different FRET levels, and then extracting rate constants using dwell-time analysis. More complex analyses have been developed, for example, using Hidden Markov Modeling (McKinney et al. 2006), to uncover information about the FRET and interconversion characteristics of states in complex systems.

1.1.2.2 Confocal Imaging/Spectroscopy

Confocal microscopy is also quite widely used in single molecule fluorescence studies. The instrumental setup (Figure 1.9) is quite similar to that used for TIRF.

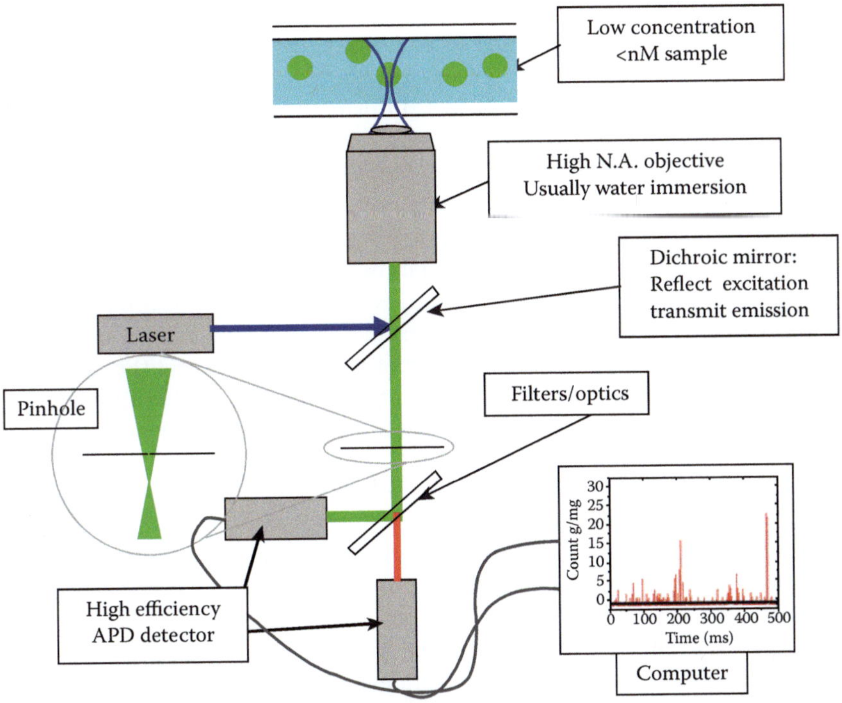

Figure 1.9 Basic instrumental setup for confocal single molecule experiments.

Here, laser illumination is via a high numerical aperture (N.A.) objective, which focuses the beam to a diffraction-limited spot. Fluorescence is collected by the same objective, separated from excitation light by a dichroic mirror, and then passes through a pinhole that further narrows the detection volume, primarily in the *Z*-dimension. The light is then treated with additional optics, for example, to split it into donor and acceptor components for FRET detection, and is then imaged onto APD detectors, which are used in a single-photon counting mode. For many applications, the number of detected photons is recorded per unit time (e.g., 1 ms or fewer), and this "intensity" is recorded as a function of time. If this is done for donor and acceptor channels, a FRET efficiency can be calculated as $E = I_a/(I_a + I_d)$, with various corrections being applied for background, cross-talk between channels, and differences in detection and emission quantum yields. A more powerful approach for recording photon events is to record detection times for each photon. From this information, intensity time-traces with arbitrarily high time resolution (down to the ~10 ns recording resolution of commonly used hardware) can be computed. In addition, correlation information can be computed from the same data. Seidel and coworkers (Sisamakis et al. 2010) have done extensive work on multiparameter fluorescence detection, where for each photon, multiple fluorescence properties are recorded, including macroscopic detection time (as above), microscopic detection time (with respect to a picosecond laser excitation pulse), polarization, and color can be simultaneously recorded using a multidetector setup. These data can then be used to recreate an intensity time-trace and measure lifetimes, properties, and FRET efficiencies for single molecules, thus proving to be a powerful advanced data collection method.

Confocal data collection can be carried out on immobilized molecules as for TIRF, but is often also used to detect diffusing or flowing molecules. This latter type of measurement largely eliminates potential perturbation from surface immobilization and allows collection of molecular properties for large numbers of single molecules in a relatively short time (Mukhopadhyay and Deniz 2007; Deniz et al. 2008). On the contrary, long time-traces cannot be obtained by this method. Data for immobilized molecules in this format are usually acquired by scanning the sample using a piezoelectric or other scanner. Scanning is used to find single molecules and position the excitation/detection volume over the single molecule, collect photons until photobleaching, then repeat the search/positioning. Time trajectories of fluorescence signals are collected in this manner and can be analyzed by methods as discussed in the TIRF section in Section 1.1.2.

1.1.2.3 *Attachment of Fluorescent Dyes for Single Molecule Fluorescence Experiments*

Single molecule fluorescence experiments usually require the use of bright externally introduced dyes. These dyes need to be attached at a specific location in the molecule, thus providing a specific probe for the properties of the molecule. Several good chemistries exist for this purpose (Ferreon et al. 2010).

For short DNA and RNA oligonucleotides that can be made synthetically (commercially available), many dyes can be attached during synthesis, hugely simplifying this process. Alternatively, primary amine or thiol functionalities may be attached, which permit specific reactions of dyes with *N*-hydroxysuccinimide or maleimide reactive groups (also available commercially for many common dyes). Both of these possibilities have been extensively utilized in single molecule work on these molecules. Larger DNA may be labeled by clipping out a piece (using a restriction endonuclease), then ligating in a smaller piece that was labeled using one of the above methods. Labeling larger RNA molecules is more difficult. An interesting method for labeling involves enlarging loops on the RNA at strategic locations, then hybridizing a short prelabeled DNA oligonucleotide to this loop (Dorywalska et al. 2005). This method was used for labeling positions on ribosomal RNA, resulting in neat data on its motion during function.

Protein labeling can range from relatively simple to very difficult (Ferreon et al. 2010). Many single molecule experiments to date have been carried out on small proteins that are either naturally cysteine-free or can be made cysteine-free without loss of structure and function. Since cysteine chemistry can be made orthogonal to the chemistries of the other amino acids, introduction of one or more cysteines at specific locations in the protein then allows site-specific labeling, usually via maleimide-functionalized dye. More complex labeling chemistries and techniques have also been used including amine labeling (generally hard to control specificity due to the presence of a large number of amines), synthetic preparation, and if needed, ligation (Deniz et al. 2000), introduction of unnatural amino acids with orthogonal chemistries (Brustad et al. 2008), and use of enzyme-mediated chemistry (Jager et al. 2006). Labeling proteins for FRET measurements is usually more complex because both donor and acceptor dyes must be attached site specifically. Simpler labeling schemes have used two cysteines for labeling. More complex and specific methods have used combinations of the techniques outlined above (Jager et al. 2006; Brustad et al. 2008). These were prepared by use of the above or similar methods. It should be noted that this is sometimes a nontrivial and time-consuming component of single molecule fluorescence experiments that should be taken into account when planning experiments on systems of interest.

1.2 Surface Attachment of Biomolecules

Many single molecule studies require that the molecule of interest be immobilized on a surface in order to track its properties for an extended period of time. In fluorescence experiments, many of the surfaces are glass or quartz (fused silica). Both noncovalent and covalent immobilization chemistries have been used, for example, noncovalent attachment of biotin-functionalized bovine serum albumin on a glass surface, followed by binding of streptavidin. This presents a surface with exposed biotin binding sites (from the multivalent streptavidin), and biotin derivatized molecules can be bound to this surface. For example, early studies of RNA folding by single molecule fluorescence used this relatively simple method. Many studies have used attachment of polyethylene glycol (PEG) via silanes (or silatranes [Lv et al. 2015]) to the surfaces, with a small fraction of biotinylated or chemically activated species for attachment of biomolecules. The PEG surface results in much lower nonspecific sticking, which is important for single molecule studies.

1.3 Some Key Examples

We now discuss some interesting examples of the use of single molecule methods in biophysical and biological studies. Some examples have also been briefly discussed during the earlier sections of this chapter.

1.3.1 Enzyme Function

How enzymes perform their functions is a topic of enduring interest from the perspective of the mechanisms and also in order to understand how to design enzymes with improved or novel function. Single molecule methods can help in such an understanding by providing detailed information about different steps in the enzymatic process and how structural properties of the enzyme influence the chemistry. A broad range of single molecule tools are useful for this purpose, including fluorescence of cofactors or substrates (which can signal chemical changes) and smFRET (which can report on conformational properties). Early studies in this area by Yeung (Xue and Yeung 1995) and Dovichi (Craig et al. 1996) studied single enzymes (results averaged over multiple turnovers) using fluorogenic substrates. A major result was that different enzymes showed different enzymatic rates. In 1998, Lu and Xie published a landmark paper (Lu et al. 1998), where they followed individual turnovers of the enzyme cholesterol oxidase, using imaging of the proteins within a gel. Their results showed that individual enzymatic rates varied not only between different molecules (termed *static heterogeneity*) but also for a single molecule as a function of time (dynamic heterogeneity). Furthermore, the variations in time showed a memory effect, which they ascribed to the protein accessing different conformational substrates. Another important example was the first single molecule study of the rotary motor protein F1-ATPase. Structural and biochemical studies indicated that this protein is a rotary motor, and these studies garnered a Nobel Prize in Chemistry in 1997. The same year, the group of Kinosita used single molecule imaging to directly monitor the rotation of the central subunit of this protein, directly visualizing the rotary nature of this protein (Noji et al. 1997). This landmark paper stimulated a series of studies on this and other systems, yielding a wealth of detailed biophysical and biochemical information about enzyme function.

1.3.2 Structural Biophysics

The conformational properties of individual molecules as well as their intricate organization into complex molecular machines both are key features of cellular function. An array of different tools has been developed to interrogate these aspects, which extend more conventional static structural biology into a realm of structural distributions and dynamics. For example, close examination reveals complexity in many folding reactions, and single molecule fluorescence and manipulation methods can uncover this complexity in terms of conformational distributions, folding and assembly intermediates, and parallel pathways. One recent example is the use of single molecule fluorescence by Benke et al. to study the assembly pathways of the bacterial toxin ClyA, where the authors uncovered a complex, multipathway process leading to pores (Benke et al. 2015).

An area of particular interest in the past decade has been the application of single molecule methods to understand the structural dynamics of intrinsically disordered proteins (IDPs) (Forman-Kay and Mittag 2013; van der Lee et al. 2014; Wright and Dyson 2015). In contrast with a conventional view of proteins as having relatively well-defined three-dimensional (3D) structures, IDPs are dynamic species that populate a wide range of conformations. Several single molecule fluorescence studies have been carried out on these systems, probing their conformational distributions, complex landscapes of coupled folding and binding, and elements of allostery upon binding to multiple partners (Figure 1.10). For example, one set of studies has focused on studies of the folding and aggregation of the amyloidogenic and Parkinson disease—linked protein α-synuclein. Because membrane binding is thought to be important for its function and also can modulate its aggregation propensity, its conformational changes upon binding have been the subject of a number of studies over the past decade. This binding-coupled-folding also has more general implications, since several IDPs show this characteristic, and it is often important for the functions in cells. While there was some debate about specific structures that were formed upon binding of α-synuclein to membranes or mimics, smFRET studies by Ferreon et al. clearly showed that a complex multistate binding-folding landscape was populated by this two-component interaction system, with multiple helical structures being populated, sometimes under one set of conditions (Figure 1.10a) (Ferreon et al. 2009). Another study used single molecule fluorescence methods and detected the presence of two types of dimers for this protein, with different biophysical characteristics and which showed modulation with Parkinson disease–related mutations (Lv et al. 2015). Additional binding components substantially increase the state complexity of relatively simple IDPs. For example, Ferreon et al. carried out smFRET studies on a ternary interaction

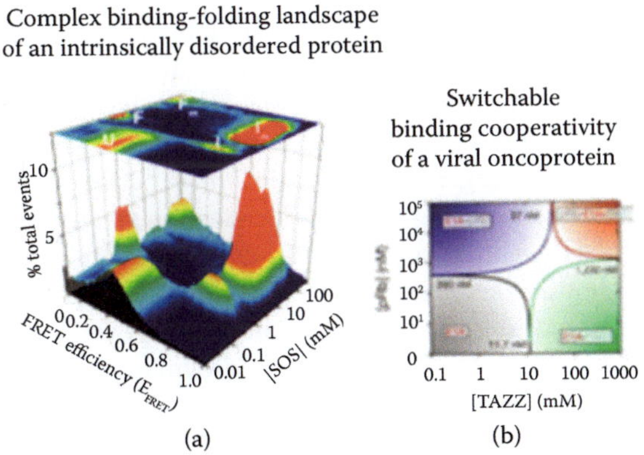

Figure 1.10 Selected applications of single molecule FRET. (a) The complex binding-folding landscape of an amyloidogenic and intrinsically disordered protein α-synuclein by smFRET. (Adapted from Ferreon, A. C., et al., *Proc. Natl. Acad. Sci. U S A*, 106, 5645, 2009.) (b) Binding phase diagrams from smFRET reveal switchable cooperativity in a ternary interaction system with the viral oncoprotein E1A. (Adapted from Ferreon, A. C., et al., *Nature*, 498, 390, 2013.)

system, with one component being the viral oncoprotein E1A. The work revealed that interaction elements in E1A could modulate binding to two biologically important partners, strikingly even reversing the sign of the cooperativity in one case. Since the cooperativity can alter the occupation probabilities of the binary and ternary complexes, and these complexes have different cellular functions, the work illustrates how complex patterns of structure and regulation could emerge from seemingly simple IDP interaction systems.

Another interesting area that has seen a lot of work is that of motor and DNA-associated proteins moving along their tracks, which are important in many cellular functions. We briefly mentioned the work of Kinosita and coworkers on the rotary motor F1-ATPase (Noji et al. 1997). In this study, the authors immobilized individual ATPase molecules and directly recorded their movement using fluorescence imaging. Further studies directly revealed the rotary movement of this protein and complex stepping patterns that accompanied its biochemical functions. A more recent example is the study by Myong et al. (2005), where they studied the movement of Rep helicase on DNA. Their interesting observation was that the protein moved along single-stranded DNA until it ran into double-stranded DNA, whereupon it dissociated and restarted the cycle. The study is also noteworthy because it used an interesting fluorescence enhancement induced by proximity of the protein. Recent photophysical studies have provided support for a mechanism for the enhancement being a reduction in *cis-trans* isomerization of the fluorescent dye due to steric interactions with the protein (Stennett et al. 2015). More recent studies have used DNA curtains, and observed what happens when proteins moving along DNA encounter other proteins, with several outcomes observed that may be particularly relevant for the situation in a cell (Lee et al. 2014).

1.3.3 Molecular Structure, Binding, and Function in Cells

Progress in biophysics needs progress in both highly reductionist approaches that allow in-depth investigations under controlled conditions, as well as investigations in more natural environments, where control is harder, but results are more relevant to the biology. Single molecule methods can contribute immensely in this area, too, although the challenges are greater than for *in vitro* experiments. Thus, fluorescence has been used for understanding details of localization and movement of molecules and species in cells or cellular contexts. Jain et al. used single molecule pull-down experiments to capture the helicase PcrA with immobilized DNA, and were able to record shuttling motions of the helicase on the DNA (Jain et al. 2011). Exciting recent advances have revealed the statistics and dynamics of the birth and movement of proteins and RNAs in live cells (Taniguchi et al. 2010; Larson et al. 2011) and even in tissues (Lyubimova et al. 2013). Recent work has also directly studied protein folding using smFRET in live eukaryotic cells (Konig et al. 2015), showing the ability to use sophisticated single molecule fluorescence methods in this context.

1.4 Concluding Remarks

Over the past almost three decades, single molecule fluorescence biophysics methods have revolutionized the way we do many experiments to probe the biophysics of proteins, DNA, and RNA, and their interactions both *in vitro* and *in vivo*. While many techniques are quite well established, studies continue to push resolution in many dimensions and also extend the techniques to new and more complex environments such as cells. The recent work directly measuring protein conformational properties and dynamics in cells is an important case in point (Konig et al. 2015). Such studies will be critical to understand complex folding, interaction, and functional landscapes of biomolecules, as well as how they compare with *in vitro* findings. An important dimension is to further understand the properties of multiple single molecules as they interact with each other, which may reveal emergent behavior that is relevant to cells. Finally, it is anticipated that high-resolution single molecule studies will continue to be extended further to live cells, tissues, and organisms.

References

Bates, M., Huang, B. & Zhuang, X. 2008. Super-resolution microscopy by nanoscale localization of photo-switchable fluorescent probes. *Curr Opin Chem Biol*, 12, 505–14.

Benke, S., Roderer, D., Wunderlich, B., Nettels, D., Glockshuber, R. & Schuler, B. 2015. The assembly dynamics of the cytolytic pore toxin. ClyA. *Nat Commun*, 6, 6198.

Betzig, E., Patterson, G. H., Sougrat, R., Lindwasser, O. W., Olenych, S., Bonifacino, J. S., Davidson, M. W., Lippincott-Schwartz, J. & Hess, H. F. 2006. Imaging intracellular fluorescent proteins at nanometer resolution. *Science*, 313, 1642–5.

Brandenburg, B. & Zhuang, X. 2007. Virus trafficking—Learning from single-virus tracking. *Nat Rev Microbiol*, 5, 197–208.

Brustad, E. M., Lemke, E. A., Schultz, P. G. & Deniz, A. A. 2008. A general and efficient method for the site-specific dual-labeling of proteins for single molecule fluorescence resonance energy transfer. *J Am Chem Soc*, 130, 17664–5.

Chang, Y. P., Pinaud, F., Antelman, J. & Weiss, S. 2008. Tracking bio-molecules in live cells using quantum dots. *J Biophotonics*, 1, 287–98.

Clamme, J. P. & Deniz, A. A. 2005. Three-color single-molecule fluorescence resonance energy transfer. *Chemphyschem*, 6, 74–7.

Craig, D. B., Arriaga, E. A., Wong, J. C. Y., Lu, H. & Dovichi, N. J. 1996. Studies on single alkaline phosphatase molecules: Reaction rate and activation energy of a reaction catalyzed by a single molecule and the effect of thermal denaturation—The death of an enzyme. *J Am Chem Soc*, 118, 5245–53.

Deniz, A. A., Dahan, M., Grunwell, J. R., Ha, T., Faulhaber, A. E., Chemla, D. S., Weiss, S. & Schultz, P. G. 1999. Single-pair fluorescence resonance energy transfer on freely diffusing molecules: Observation of Forster distance dependence and subpopulations. *Proc Natl Acad Sci U S A*, 96, 3670–5.

Deniz, A. A., Laurence, T. A., Beligere, G. S., Dahan, M., Martin, A. B., Chemla, D. S., Dawson, P. E., Schultz, P. G. & Weiss, S. 2000. Single-molecule protein folding: Diffusion fluorescence resonance energy transfer studies of the denaturation of chymotrypsin inhibitor 2. *Proc Natl Acad Sci U S A*, 97, 5179–84.

Deniz, A. A., Mukhopadhyay, S. & Lemke, E. A. 2008. Single-molecule biophysics: At the interface of biology, physics and chemistry. *J R Soc Interface,* 5, 15–45.

Dorywalska, M., Blanchard, S. C., Gonzalez, R. L., Kim, H. D., Chu, S. & Puglisi, J. D. 2005. Site-specific labeling of the ribosome for single-molecule spectroscopy. *Nucleic Acids Res,* 33, 182–9.

Eigen, M. & Rigler, R. 1994. Sorting single molecules—Application to diagnostics and evolutionary biotechnology. *Proc Natl Acad Sciences U S A,* 91, 5740–7.

Ferreon, A. C., Gambin, Y., Lemke, E. A. & Deniz, A. A. 2009. Interplay of alpha-synuclein binding and conformational switching probed by single-molecule fluorescence. *Proc Natl Acad Sci U S A,* 106, 5645–50.

Ferreon, A. C., Moran, C. R., Gambin, Y. & Deniz, A. A. 2010. Single-molecule fluorescence studies of intrinsically disordered proteins. *Methods Enzymol,* 472, 179–204.

Forman-Kay, J. D. & Mittag, T. 2013. From sequence and forces to structure, function, and evolution of intrinsically disordered proteins. *Structure,* 21, 1492–9.

Gambin, Y. & Deniz, A. A. 2010. Multicolor single-molecule FRET to explore protein folding and binding. *Mol Biosyst,* 6(9), 1540–7.

Greenleaf, W. J., Woodside, M. T. & Block, S. M. 2007. High-resolution, single-molecule measurements of biomolecular motion. *Annu Rev Biophys Biomol Struct,* 36, 171–90.

Ha, T., Enderle, T., Ogletree, D. F., Chemla, D. S., Selvin, P. R. & Weiss, S. 1996. Probing the interaction between two single molecules: Fluorescence resonance energy transfer between a single donor and a single acceptor. *Proc Natl Acad Sci U S A,* 93, 6264–8.

Hohng, S., Joo, C. & Ha, T. 2004. Single-molecule three-color FRET. *Biophys J,* 87, 1328–37.

Jager, M., Nir, E. & Weiss, S. 2006. Site-specific labeling of proteins for single-molecule FRET by combining chemical and enzymatic modification. *Protein Sci,* 15, 640–6.

Jain, A., Liu, R., Ramani, B., Arauz, E., Ishitsuka, Y., Ragunathan, K., Park, J., Chen, J., Xiang, Y. K. & Ha, T. 2011. Probing cellular protein complexes using single-molecule pull-down. *Nature,* 473, 484–8.

Kim, S. A., Heinze, K. G. & Schwille, P. 2007. Fluorescence correlation spectroscopy in living cells. *Nat Meth,* 4, 963–73.

Konig, I., Zarrine-Afsar, A., Aznauryan, M., Soranno, A., Wunderlich, B., Dingfelder, F., Stuber, J. C., Pluckthun, A., Nettels, D. & Schuler, B. 2015. Single-molecule spectroscopy of protein conformational dynamics in live eukaryotic cells. *Nat Meth,* 12, 773–9.

Larson, D. R., Zenklusen, D., Wu, B., Chao, J. A. & Singer, R. H. 2011. Real-time observation of transcription initiation and elongation on an endogenous yeast gene. *Science,* 332, 475–8.

Lee, J. Y., Finkelstein, I. J., Arciszewska, L. K., Sherratt, D. J. & Greene, E. C. 2014. Single-molecule imaging of FtsK translocation reveals mechanistic features of protein-protein collisions on DNA. *Mol Cell,* 54, 832–43.

Lu, H. P., Xun, L. & Xie, X. S. 1998. Single-molecule enzymatic dynamics. *Science,* 282, 1877–82.

Lv, Z., Krasnoslobodtsev, A. V., Zhang, Y., Ysselstein, D., Rochet, J. C., Blanchard, S. C. & Lyubchenko, Y. L. 2015. Direct detection of alpha-Synuclein dimerization dynamics: Single-molecule fluorescence analysis. *Biophys J,* 108, 2038–47.

Lyubimova, A., Itzkovitz, S., Junker, J. P., Fan, Z. P., Wu, X. & van Oudenaarden, A. 2013. Single-molecule mRNA detection and counting in mammalian tissue. *Nat Protoc,* 8, 1743–58.

Magde, D., Webb, W. W. & Elson, E. 1972. Thermodynamic fluctuations in a reacting system—Measurement by fluorescence correlation spectroscopy. *Phys Rev Lett,* 29, 705.

McKinney, S. A., Joo, C. & Ha, T. 2006. Analysis of single-molecule FRET trajectories using hidden Markov modeling. *Biophys J,* 91, 1941–51.

Mukhopadhyay, S. & Deniz, A. A. 2007. Fluorescence from diffusing single molecules illuminates biomolecular structure and dynamics. *J Fluoresc,* 17, 775–83.

Mukhopadhyay, S., Krishnan, R., Lemke, E. A., Lindquist, S. & Deniz, A. A. 2007. A natively unfolded yeast prion monomer adopts an ensemble of collapsed and rapidly fluctuating structures. *Proc Natl Acad Sci U S A,* 104, 2649–54.

Myong, S., Rasnik, I., Joo, C., Lohman, T. M. & Ha, T. 2005. Repetitive shuttling of a motor protein on DNA. *Nature,* 437, 1321–5.

Noji, H., Yasuda, R., Yoshida, M. & Kinosita, K., Jr. 1997. Direct observation of the rotation of F1-ATPase. *Nature,* 386, 299–302.

Patterson, G., Davidson, M., Manley, S. & Lippincott-Schwartz, J. 2010. Superresolution imaging using single-molecule localization. *Annu Rev Phys Chem,* 61, 345–67.

Rigler, R., Mets, U., Widengren, J. & Kask, P. 1993. Fluorescence correlation spectroscopy with high count rate and low-background—Analysis of translational diffusion. *Eur Biophys J Biophys Lett,* 22, 169–75.

Rust, M. J., Bates, M. & Zhuang, X. 2006. Sub-diffraction-limit imaging by stochastic optical reconstruction microscopy (STORM). *Nat Meth,* 3, 793–5.

Sisamakis, E., Valeri, A., Kalinin, S. Rothwell, P. J. & Seidel, C. A. 2010. Accurate single-molecule FRET studies using multiparameter fluorescence detection. *Methods Enzymol.* 475, 455–514.

Stennett, E. M. S., Ciuba, M. A., Lin, S. & Levitus, M. 2015. Demystifying PIFE: The photophysics behind the protein-induced fluorescence enhancement phenomenon in Cy3. *J. Phys. Chem. Lett.* 6, 1819–23.

Stryer, L. & Haugland, R. P. 1967. Energy Trasfer: A spectroscopic ruler. *Proc Natl Acad Sci U S A,* 58, 719–26.

Taniguchi, Y., Choi, P. J., Li, G. W., Chen, H., Babu, M., Hearn, J., Emili, A. & Xie, X. S. 2010. Quantifying E. coli proteome and transcriptome with single-molecule sensitivity in single cells. *Science,* 329, 533–8.

Thompson, N. L., Lieto, A. M. & Allen, N. W. 2002. Recent advances in fluorescence correlation spectroscopy. *Curr Opin Struct Biol,* 12, 634–41.

van der Lee, R., Buljan, M., Lang, B., Weatheritt, R. J., Daughdrill, G. W., Dunker, A. K., Fuxreiter, M., et al. 2014. Classification of intrinsically disordered regions and proteins. *Chem Rev,* 114, 6589–631.

Wright, P. E. & Dyson, H. J. 2015. Intrinsically disordered proteins in cellular signalling and regulation. *Nat Rev Mol Cell Biol,* 16, 18–29.

Xue, Q. & Yeung, E. S. 1995. Differences in the chemical reactivity of individual molecules of an enzyme. *Nature,* 373, 681–3.

2 Atomic Force Microscope for Topographic Studies

Yuri L. Lyubchenko
University of Nebraska Medical Center
Omaha, Nebraska

Contents

The atomic force microscope (AFM), also known as a scanning force microscope (SFM), belongs to a family of microscopes termed *scanning probe microscopes* (SPMs). Unlike optical and other microscopes that utilize diffraction phenomenon primarily of the electromagnetic field, SPM instruments provide topographic images of surfaces by scanning them with a sharp probe. Having been invented in the early 1980s, SPM is a novel family of microscopes that started with the invention in 1982 by Binnig and Rohrer of the scanning tunneling microscope (STM) (Binnig 1982). This invention brought a revolution to the surface sciences due to its capability to visualize individual atoms on a crystal surface. The inventors were awarded a Nobel Prize in physics in 1986 for their fundamental discovery; interestingly, in that same year, one of the Nobel Prize winners, G. Binnig, published a paper entitled "Atomic Force Microscope" in which the AFM was described for the first time (Binnig 1966).

2.1 Principles of the Atomic Force Microscope

A schematic illustrating the principles of AFM operation is shown in Figure 2.1. A sharp stylus (AFM tip shown as a blue triangle) reads the topographic profile of a sample (shown as a green bumpy profile) by scanning over the sample in a raster pattern. The tip is attached to a cantilever that works as a spring, pressing the tip against the sample as it is jostled up and down following the surface profile. Note that in this particular scheme the tip is stationary, but the sample mounted on the piezoelectric scanner moves. There are a few important features of the AFM design illustrated in Figure 2.1 that are worth emphasizing: First, the position of the sample relative to the tip is controlled by the scanner, which can be done with accuracy better than 1 nm. Second, the tip can be atomically sharp.

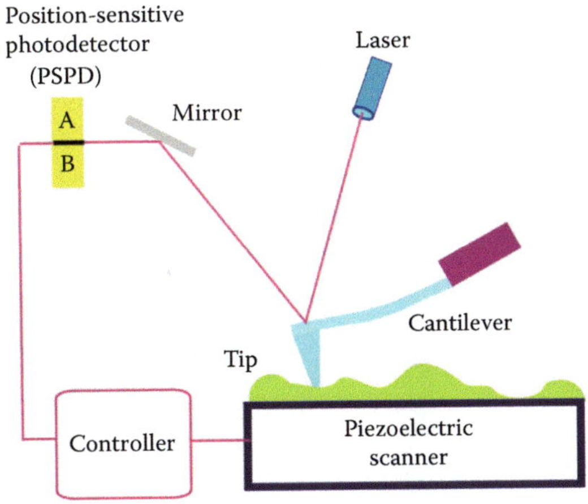

Figure 2.1 AFM.

Third, the vertical displacement of the tip relative to the surface is determined with subnanometer accuracy. These three major factors lay the foundation for the capability of AFM to provide a topographic image with atomic accuracy. The atomic resolution possible with AFM has been achieved in a number of laboratories. We mention here the paper of Sokolov et al. (1999) in which both calcium and oxygen atoms at the cleavage plane of a $CaSO_4$ surface were identified. Importantly, the images were obtained while scanning was performed in water with the use of a commercial instrument. That publication has an extensive bibliography on the AFM imaging with atomic resolution. The following paragraphs provide details on the operation of an AFM, which will aid in understanding major features of this key nanoimaging instrument.

2.1.1 Scanning Procedure

First, the position of the sample relative to the tip (or vice versa) should be controlled with nanometer accuracy or better. This requirement is achieved by the use of sample transducers or scanners made of piezoelectric materials, typically piezoceramics such as lead zirconate titanate (PZT). Piezo materials change in size depending on the voltage applied. If the electric field is parallel to the polarization vector of the material, the transducer expands. But, it contracts if the electric field is perpendicular to the materials polarization vector. This voltage-dependent change of the PZT size is schematically shown in Figure 2.2. Imagine that a sample stage is placed between the two pieces of PZT, and apply to them the same voltage but in different polarity, so the left transducer compresses, and the right one expands. As a result, the sample stage will move to the left by the value Δx. If we reverse the voltage direction applying expansion voltage to the left transducer and the same voltage of the opposite sign to the right transducer, the sample moves to the right on the same value but with a negative sign, $-\Delta x$. The value of the displacement of the sample depends on the voltage applied, and the accuracy of the sample movement is in the subnanometer range for the total scale of the displacement in the

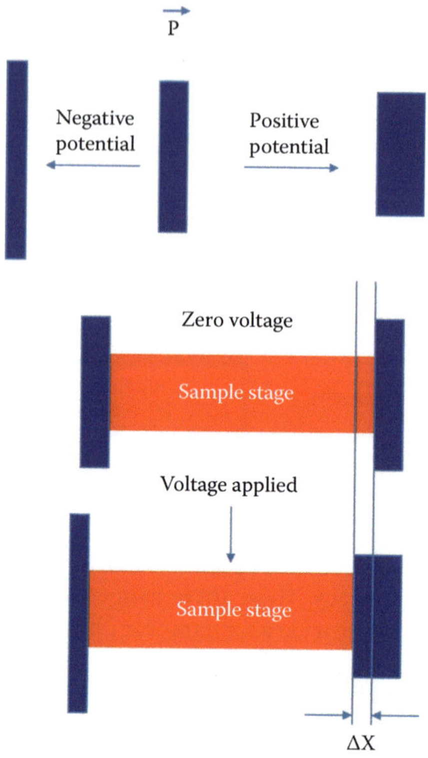

Figure 2.2 Principle of operation of the piezoscanner.

range more than 1 micron (1,000 nm). If the voltage applied to the transducers changes gradually over time, the sample will move gradually along the *X* axis as well. Similarly, if transducers are positioned at the other sides of the sample, applying the voltage allows movement of the sample along the *Y* axis.

Scanning of the sample is typically performed in the following way (see Figure 2.3). A voltage linearly changing with time is applied to *X* transducer with the highest value corresponding to the movement of the sample to a selected position X_f (red horizontal line). At this point the voltage is reversed to allow the sample to move to the original starting position X_s (blue horizontal line above the red one). A small voltage pulse is applied to the *Y* transducer pushing the sample a small step up (a green vertical line in the figure), and the same linearly increased voltage is applied to the *X* transducer to move the sample over the same distance. The voltage is reversed again. The sample is returned to the initial position; the same voltage pulse is applied to *Y* transducer, and the scanning process is repeated many times until with such a raster movement all selected area defined by Y_f value is scanned. After that the sample is returned to the initial position (X_s,Y_s) by rapidly dropping the voltage on the *Y* transducer to the initial value.

Figure 2.4 shows the pattern of the voltage applied to both transducers to accomplish the scanning process. The voltage applied to the *X* transducer has a

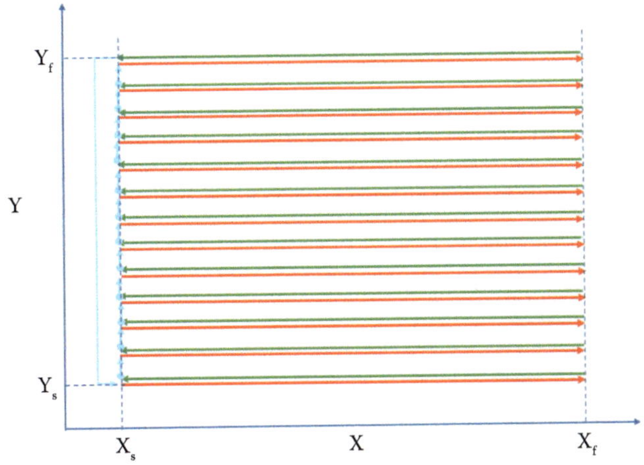

Figure 2.3 The AFM scanning process. Traces and retraces are shown as horizontal red and green lines, respectively. After completion of the scan, the sample is returned to the initial position. Jumps along the *Y*-axis are shown in blue arrows.

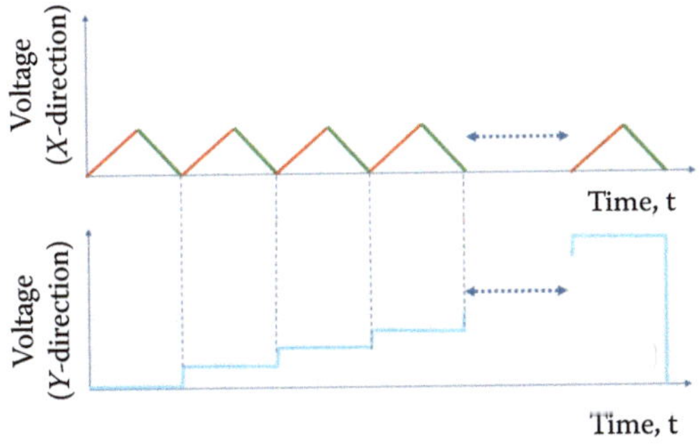

Figure 2.4 Profiles of voltages applied to *X* (top) and *Y* transducers (bottom) to provide the scanning pattern shown in Figure 2.3.

sawtooth pattern in which the amplitude of the voltage defines the range of scanning, or X_f value. The slopes of the tooth define how fast the scanning process is. The smaller the tooth base, the faster is the scanning. The voltage profile of the *Y* transducer has a stepwise pattern with the step size defining the number of horizontal scanning lines. It is critical for the entire scanning process that the movement of *X* and *Y* transducers is synchronized in such a way that the pulse to the *Y* transducer is applied exactly at the time that the retrace scan is completed. High-precision electronics are required for such synchronized operation of the scanner. See Box 2.1 for the description of the cylinder-type AFM scanner.

To detect changes in the vertical movement of the tip during scanning, a number of methods can be employed. Here we describe the most efficient and simple one in which a laser beam bounces off the back of the cantilever onto a position-sensitive photodetector (PSPD; Figure 2.1). The principle of this method is schematically shown in Figure 2.5. If the surface is flat, the laser spot equally illuminates the top and bottom detectors (Figure 2.5a). As the tip hits a bump on the surface, the cantilever bends up, the position of the laser beam on the detector shifts upward illuminating more the top detector (Figure 2.5b). On the contrary, when the tip moves down into a hole in the surface (Figure 2.5c), the cantilever bends inward, and so more light comes to the bottom detector. The PSPD itself can measure displacements of light as small as 1 nm (Shlyakhtenko et al. 2012).

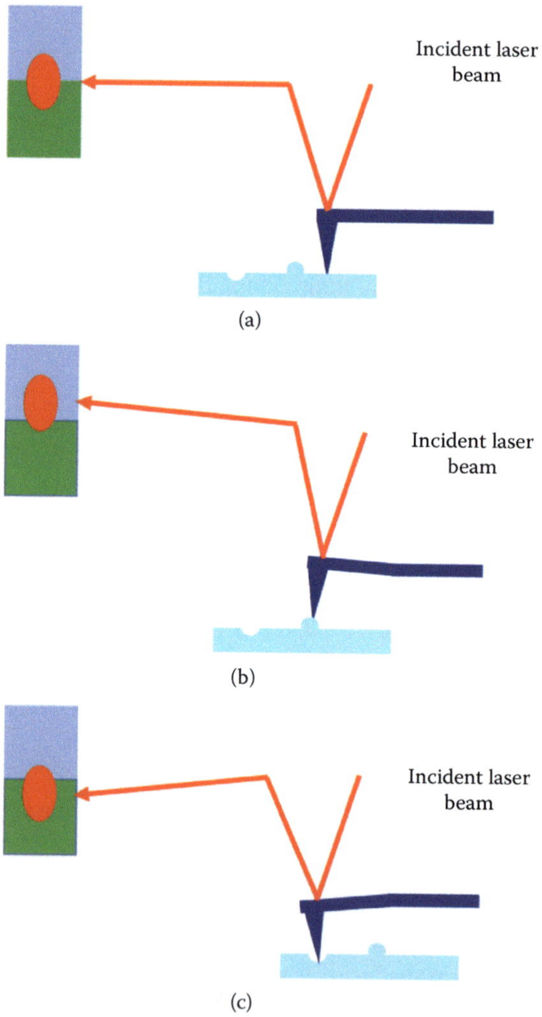

Figure 2.5 The optical detection principle for AFM. (a) The tip moves over a flat surface; (b) the tip hits a bump on the surface, the cantilever bends up, the position of the laser spot shifts upward illuminating more the top detector; (c) the tip moves to the hoe, the cantilever bends inward, and so the light spot moves to the bottom detector.

Box 2.1 Cylinder-Type AFM Scanner

The AFM scanners are made in different designs, and generally they can be divided into two major categories. In one category, *X*, *Y*, and *Z* transducers are combined in one scanner that moves either the sample as we described above or the AFM tip. Another category includes the designs in which *X* and *Y* transducers are combined in one unit separate from the *Z* transducer. One of the first types of designs has a tubular shape (Figure 2.6) in which *X*, *Y*, and *Z* transducers are indicated in different colors. Each transducer comes in pairs allowing for voltage to be applied in opposite polarities. If +V voltage is applied to the left *X* transducer it induces expansion of the left side of the tube. The same voltage but in opposite polarity is applied to the right *X*-transducer to cause the compression of the right side. As a result of this combined action of the transducers, the cylinder bends to the right causing the right-side movement of the sample mounted on the top. The change of the voltage's polarity leads to the movement of the sample to the left. Similarly, the movement along the *Y* axis is achieved by applying the voltage to the pair of *Y* transducers. Thus, applying the voltage in the pattern shown in Figure 2.4 enables scanning of the sample in a raster pattern. The central tube is the *Z*-transducer, which acts to move the sample in the vertical direction. The advantage of the one-scanner design is that it allows for the entire design of the AFM to be miniaturized. Another convenience of a tubular scanner is its versatility in terms of scanning the sample or the AFM tip. Both designs are rather equally employed by the manufacturers. The nonlinearity is the major disadvantage of tubular scanners that requires performing calibration tests on a regular basis. Specifics to this and other designs of *X*, *Y*, and *Z* scanners can be found in the literature (Mironov 2004). The nonlinearity problem is less severe in the designs in which the *X*–*Y* scanner is physically separated. Typically in such designs, the sample is mounted on the *X*–*Y* scanner, whereas the tip is mounted on the *Z*-scanner.

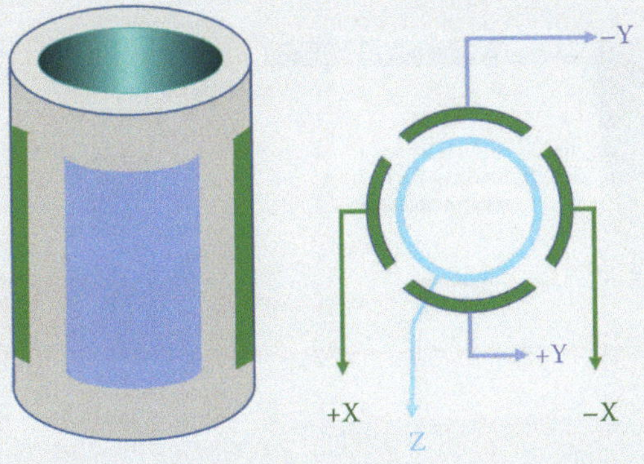

Figure 2.6 Cylinder shape AFM scanner.

However, the setup shown in Figure 2.5 allows for an increase in this sensitivity. This is due to almost two orders of magnitude difference between the lengths of the optical path of the beam reflected from the cantilever to the detector, and the cantilever size. Indeed, typical sizes of cantilevers are ~100 μm (0.1 mm), whereas the optical path is ≈1 cm (the low estimate), therefore, producing an amplification coefficient of 100. As a result, the system can detect a vertical movement of the cantilever tip as small as 0.01 nm, that is in the range of subatom sizes, providing the AFM with the capability of detecting surface structure with atomic resolution. The detection systems work in such a way that the difference between the signals from each detector is amplified by the electronics that send an appropriate voltage to the Z-piezo transducer, moving the sample up or down to reach an equal signal on both PSD detectors. The value of compensating voltage as well as signals to *X* and *Y* transducers are translated by the electronics and the software into the surface topography appearing on the computer screen during the scanning process.

2.2 What Does AFM Measure?

The schematic in Figure 2.7 explains the physical principles of AFM operation. The AFM tip (Figure 2.7a), shown as a triangle, touches the surface to generate a topographic image. Figure 2.7b illustrates the tip-sample contact at the atomic level. Red balls in this cross section are representing atoms of the surface shown in a regular two-dimensional array, and atoms of the tip are shown as blue balls. The interaction between the atoms of the tip and the surface provides the mechanism for the image reconstruction in the AFM. Let's consider the interaction between the closest atoms of the tip and the surface indicated in this scheme as an arrow. The interatom interactions are van der Waals interactions that were analytically described by Lennard-Jones potentials (Lennard-Jones 1931) that include an attraction potential that decays with the interatomic distance r as r^{-6} and repulsion potential that is much steeper than the attraction potential decaying as r^{-12}.

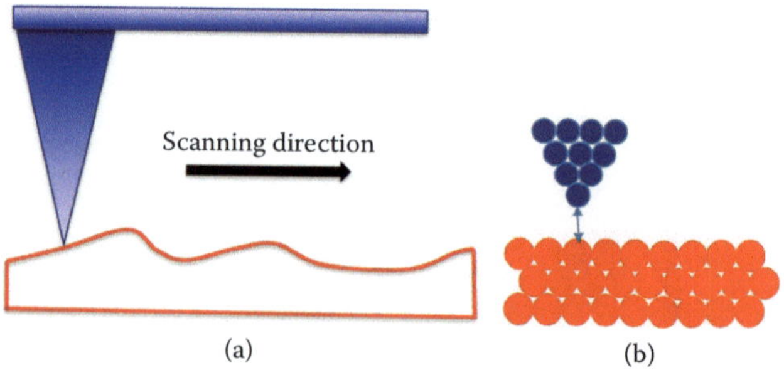

Figure 2.7 Interaction of the AFM tip with the surface atoms. (a) The scanning tip over the sample surface; (b) the same system but zoomed to the atomic level. Red balls are the atoms of the surface; the atoms of the tip are shown as blue balls.

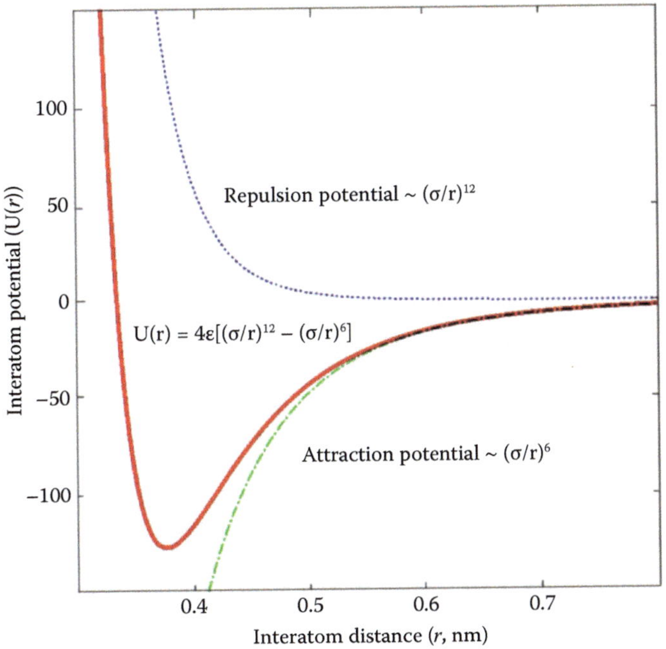

Figure 2.8 Lennard-Jones potentials for interatomic interactions. The red curve shows the dependence of the interatomic energy on the distance between the centers of the atoms ($r = |r_1 - r_2|$). Dotted curves correspond to the two components of the entire potential profile, repulsion (blue) and attraction (green).

Potential curves demonstrating interatom interactions are shown in Figure 2.8, with the attraction and repulsion sections drawn in different colors. Initially the two atoms are separated by a large distance, and the energy of the two-atom system is close to zero. As the atoms are brought together, the energy decreases due to their attraction. However, as the distance between the atoms decreases, the repulsion part of the interaction potential (blue curve) increases as observed on the potential energy profile by deviation of the red curve from the green one. The effect is dramatic at short interatomic distances at which the electron clouds of the atoms repel each other strongly, so the profile of the potential curve changes its sign, and the energy increases abruptly with further approaching of the atoms.

The described above van der Waal potential is valid for diatomic, so more realistic consideration of the interactions between the AFM tip and the sample requires another theoretical model that in addition to the closest atoms between the tip and surface incorporates contributions from other atoms of the tip and sample surfaces. This problem was considered for the first time by H. Hamaker (1937) who integrated the interactions of interatomic interactions over all atoms in the system. The analysis for the interaction of spherical particles with radius R with a flat surface in the framework of the Hamaker theory produces the following simple equation for the attraction potential as a function of the distance, D, between the sphere with radius R and the surface:

$$U = HR/(12D) \tag{2.1}$$

where H is the Hamaker constant that depends on the materials of the tip and the surface and is of the order of 10^{-19} J or 0.1 aJ (attojoule). A more rigorous analysis of the van der Waals interaction of solid bodies was performed by E. Lifshitz who showed that the Hamaker constant also includes characteristics of the medium between the interaction bodies (Dzyaloshinskii et al. 1961). An important prediction of the Lifshitz theory is that the van der Waals interaction can be reduced to very low values or even completely eliminated by the use of appropriate immersion liquids. We return to this issue later when we discuss various modes of the AFM applications, the time-lapse imaging of molecular dynamics in particular.

The Lifshitz theory was very useful in calculating the tip-sample interaction, and Figure 2.9 shows the plot made with the use of the parameters taken from the book of D. Sarid (2007). In this example the tip is approximated by a sphere with a radius of 10 nm. Note that the majority of commercially available AFM tips have a radius of curvature in the range of 10 nm, although sharper tips are available from some vendors. The full equation for the potential used for the graph plotting is given in the figure, and the necessary parameters required for the calculation of the potential are provided in the figure caption. This graph shows that the attraction between the sphere and the surface increases in short-range distances

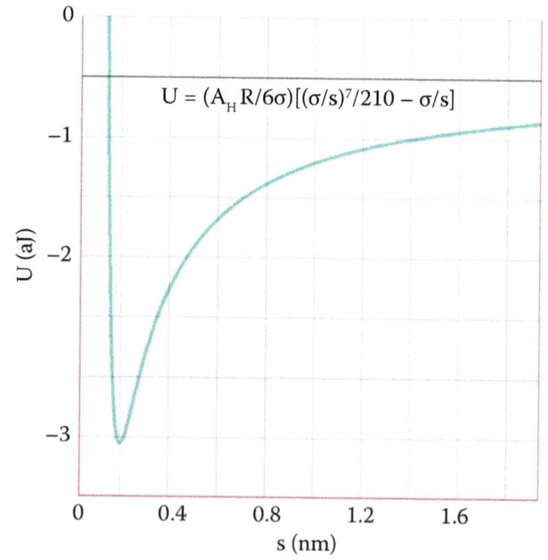

Figure 2.9 Lennard-Jones potential calculated for the tip-sample distance "s" for the following parameters of the system: tip radius R = 10 nm; interatomic distance s = 0.35 nm (red spheres in Figure 2.7b), and Hamaker constant A_H = 0.425 aJ.

(between 1.5 and 0.2 nm), reaching the maximum magnitude (minimum of the potential) –U = 3.06 aJ at ~0.19 nm, and after that the repulsion increases sharply.

2.3 AFM Resolution

It is intuitively clear that the tip geometry contributes to the AFM resolution. We follow the findings of Bustamante and colleagues (1997) to illustrate graphically the resolution principle for AFM using the scheme shown in Figure 2.10. In this figure, the tip of a smooth shape with the radius of curvature *R* at the apex scans over the sample that presents very narrow vertical bars with height *H*. As a result of scanning over the sample, the tip generates a profile that is very different from the bar-type sample, and the profile is shown in the scheme by the dotted line. This line is an AFM image of the sample that reproduces the tip profile. This effect is termed the tip convolution effect. AFM images are complex combinations of the sample and tip geometries. It is clear from this scheme that a sharp tip is required for obtaining images with a shape close to the sample shape. We use this schematic to understand the basics of AFM resolution. If we place another bar at distance *d* from the first one, the AFM images will be as shown by dotted lines. The Δz in this figure is the vertical distance between the object's top and the intersection of the two dotted profiles (a dimple in the profile). According to the Rayleigh criterion, the minimal distance d_{min} that allows for these profiles to be separate features on the images corresponds to $\Delta z \approx 0.2$ H. This condition leads to the following equation for d_{min}:

$$d_{min} = 2\ (0.4\ R\ H)^{0.5} \tag{2.2}$$

This equation demonstrates that a sharper tip results in a smaller d_{min}, which in turn yields a higher resolution. If *R* and *H* are equal to 0.1 nm, which corresponds to the atom size, we can obtain a d_{min} value ≈ 0.12 nm. So atomic resolution with AFM is possible provided the tip is atomically sharp. If the sample is

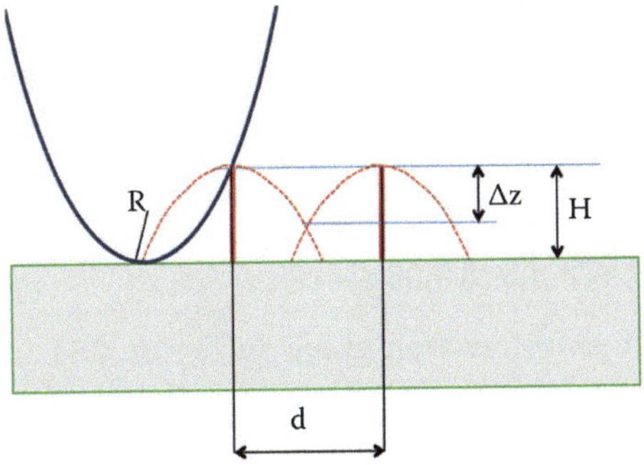

Figure 2.10 The effect of tip on the AFM resolution. See the text for explanation. (Reproduced from *Micron*, 42, Lyubchenko, Y.L., Preparation of DNA and nucleoprotein samples for AFM imaging, 196–206, 2011, Copyright © (2011), with permission from Elsevier.)

double-stranded DNA with a 2 nm diameter and the tip radius of curvature is 10 nm, which is a typical nominal value for regular AFM tips, the value $d_{min} \approx$ 5.7 nm can be obtained. Sharper tips are needed to get higher resolution, and the recent advances in fabrication of ultrasharp AFM tips (radius of curvature R ~ 1 nm) have made it possible to perform high-resolution studies of nucleic acids of different types and conformations (Klinov et al. 2007).

Although the sharpness of the tip is a major contributor to AFM resolution, there are a number of factors limiting AFM resolution, with thermal motion among them. The implementation of cryo AFM has made it possible to dramatically improve the resolution of isolated molecules to the nanometer level (Zhang et al. 1996). According to that paper, cooling the sample down to 85K resulted in resolving the domain structure of isolated immunoglobulin IgA protein in various states.

Capillary effect is another factor dramatically influencing AFM resolution. When a surface is exposed to air, it absorbs water from vapors and forms a thin water layer over the surface. Mica is one of the traditional AFM substrates, and it is very hydrophilic, retaining water molecules that can be removed in very high vacuum. AFM tips, made from either silicon or silicon nitride, are quite hydrophilic as well, so the water layer further increases the probe size, thereby decreasing the resolution. However, hydration of the sample and tip lead to another complication; upon approaching the AFM tip to the surface, a water bridge between the hydrated tip and sample is formed. The high surface tension leads to capillary forces as large as dozens of nano Newton (nN) (Freund et al. 1999). Therefore, the sample can be dragged over the surface during scanning, deteriorating the imaging process. Drying of the sample and imaging at low relative humidity help to reduce the effect of capillary forces. Alternatively, the capillary effect can be reduced if scanning is performed in vacuum, but vacuum is not attractive for imaging of biological samples. At the same time, there are no capillary forces if the imaging is performed in aqueous solution, and given the fact that imaging in aqueous solutions is the best way to preserve biological samples, imaging in liquids is very attractive for biological studies. Indeed, the images of DNA with highest resolution were obtained for images acquired in aqueous solutions (Mou et al. 1995; Lyubchenko and Shlyakhtenko 1997). Note that in other studies (Lyubchenko and Shlyakhtenko 1997), high-resolution images were obtained for individual DNA molecules rather than tightly packed DNA molecules (Mou et al. 1995). This issue is discussed in Section 2.6.

2.4 Imaging Modes for AFM

We use the tip-sample interaction graph in Figure 2.11 to illustrate the major features of AFM operation. Upon approaching the surface, the tip comes to the attraction regime in which primarily the atoms at very close proximity to the tip apex primarily contribute to the tip-sample interaction. After passing the minimum on the potential curve, the repulsion interaction rises sharply, achieving large positive values in the very short range of the tip-sample distances. This scanning regime is termed the *contact mode*. The interaction of the tip and the surface is strong, so the deformation of a soft sample typically occurs. Schematically this

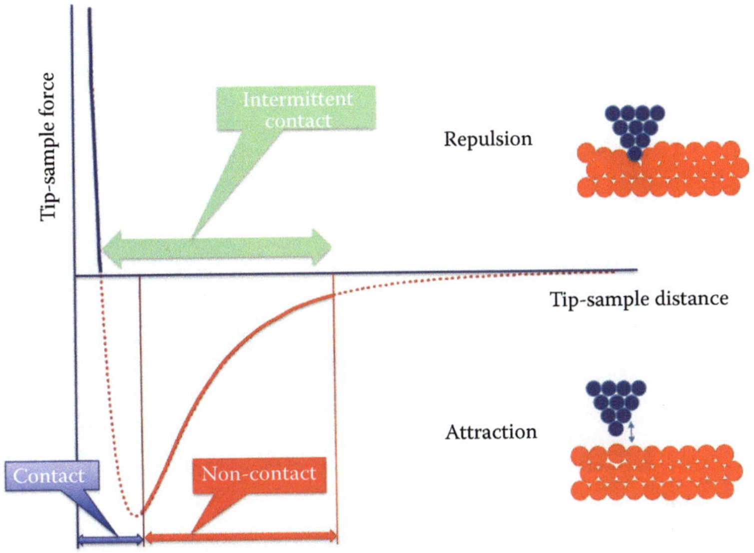

Figure 2.11 Scanning regimes of AFM.

deformation is illustrated in the top right cartoon by displacement of the red balls. For hard samples at small tip-surface separation, the pressure created by the tip at the apex is so strong that atoms can jump from the tip to the surface, and vice versa. Therefore, the contact mode in AFM operation does not allow one to obtain reliable and stable images of soft samples at high resolution. In addition, damage can be made to the tip during approach or scanning, and indeed tip crashing events are not rare if the AFM operates in contact mode.

A solution was found with the invention of the oscillating mode by Martin et al. (1987) in which a cantilever is deliberately vibrated at the frequency close to the cantilever resonant frequency by a piezoelectric modulator. As a result, the tip-sample contact is limited to a short period of time as defined by the oscillation frequency, which is ~100–300 kHz for imaging in air. This approach is currently known as the tapping mode (TM) (Zhong et al. 1993). However, this term has been trademarked by the AFM manufacturer Veeco; therefore, other manufacturers introduced this methodology under different names, such as intermittent contact (IC) and alternating contact (AC), to mention a couple. Schematically, the range of operation of IC (TM) mode is shown in Figure 2.11 with a thick horizontal double arrow.

Figure 2.12 shows images of two biological samples to illustrate the usefulness of TM for the imaging of biological samples. Frame a of the figure shows an AFM image of circular DNA with an extruded cruciform structure (indicated with an arrow) (Shlyakhtenko et al. 2000a). DNA appears as a filament with a uniform thickness with no indication of tip damage to the sample. Alternative DNA structures such as a cruciform appear in cells transiently, and their functional role is widely discussed (reviewed in Bikard et al. 2010). In addition to

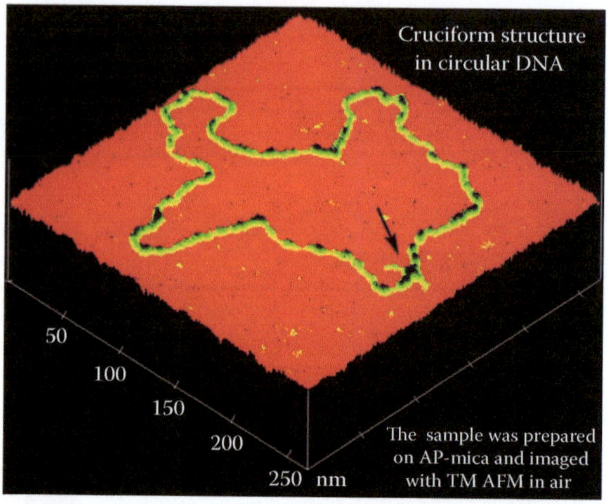

(a)

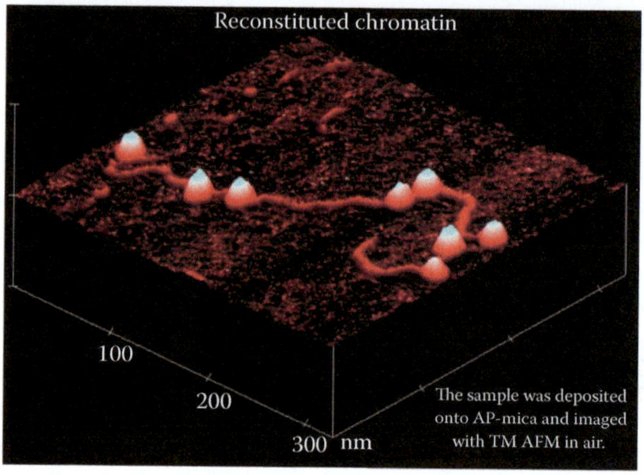

(b)

Figure 2.12 AFM images of supercoiled DNA prepared with the use of AP-mica. (a) A large-scale image with three DNA molecules in the field. (b) Zoomed image of the top-left molecule in (a). The supercoiled molecules appear as uniformly twisted filaments with a plectonemic shape. (Reproduced from *Micron*, 42, Lyubchenko, Y.L., Preparation of DNA and nucleoprotein samples for AFM imaging, 196–206, 2011, Copyright © (2011), with permission from Elsevier.)

cruciforms, unwound DNA structures and intramolecular triplexes, H-DNA, have been imaged with AFM (reviewed in Lyubchenko 2004). Note that transmission electron microscopy (TEM), which has a resolution similar to AFM, was not that efficient in structural studies of alternative DNA structures. Therefore, the ability of AFM to image alternative DNA structures makes this microscopy very attractive in the field of molecular and structural genetics.

Figure 2.12b shows the image of reconstituted chromatin assembled on purified histone octamers studied in Yodh et al. (2002). Nucleosomal particles formed by wrapping double-stranded DNA around the histone octamers are unambiguously seen as bright globular features (blobs). In most cases, the sizes of the blobs are found to be similar, and there is no apparent damage to the nucleosomes, which confirms again the appropriateness of TM imaging mode for use in the imaging of such large biomolecular complexes. Electron microscopy was initially applied to prove the beads-on-string morphology of isolated or reconstituted chromatin, but reliable imaging required a tedious sample preparation including staining of the sample. The images shown in frame b were obtained without any staining, where the sample was instead cross-linked with glutaraldehyde. Recent advances in the preparation of samples for AFM imaging made it possible to obtain images of chromatin not subjected to the cross-linking step (Lyubchenko and Shlyakhtenko 2009; Shlyakhtenko et al. 2009b).

Intrinsically, tapping mode decreases the effect of the tip-sample interaction due to the short time of tip-sample contact; however, an additional factor influencing the tip-sample interaction is tip oscillation amplitude. Indeed, the energy provided by the tip to the sample depends quadratically on the drive amplitude of the oscillating probe (Lyubchenko and Shlyakhtenko 1997). Therefore, decreasing this parameter should further preserve the sample structure. This issue was investigated thoroughly in the lab of Thomson and led to the development of the low tip oscillation mode termed *small amplitude small set-point* (SASS), which operates close to the zero (net) sample force (Santos et al. 2013). In this mode the tip oscillates at amplitudes below 5 nm, which is an order of magnitude lower than typical amplitude values for TM AFM. As a result, the authors were able to scan continuously over a DNA sample for hours without visible damage to the DNA molecules.

The tip-sample interaction can be weak if the oscillation amplitude is low enough so that the tip oscillates in the range of attractive interaction, a regime termed *noncontact mode*. However, according to Figure 2.9, the amplitude of the tip oscillation should be less than 1 nm. This requirement can be met for imaging in vacuum to eliminate the hydration effect and remove impurities, and with the use of atomically sharp tips (the tip was terminated with the C=O group), acquiring images at temperatures close to absolute zero value (Gross et al. 2012). Using this cryo-AFM instrument, the authors obtained images of organic molecules with atomic resolution. Moreover, the measurements were so precise that the difference in lengths of single and double covalent bonds was determined. It is also worth noting that in the early work (Sokolov et al. 1999) and more recent publications (e.g., Rode et al. 2009), true atomic resolution of a calcite surface was achieved using the AFM microscope operating in aqueous solutions. To accomplish this, a home-built controller was used, and a special liquid cell was designed, which together allowed for the group to keep the cantilever oscillation amplitude constant at a level of 0.2–2 nm and acquire images with amplitudes in the range of 0.2–0.5 nm. While nanometer range resolution is possible to achieve, such strict imaging conditions are difficult to maintain for the imaging of soft biological samples.

2.5 AFM Sample Preparation

Surface preparation is a key step for imaging samples with any type of microscopy to ensure reliable and reproducible data acquisition. Surface preparation requires additional attention for SPM because the scanning tip can change the sample structure and dynamics. Moreover, the tip can displace the sample, leading to what are notoriously known as surface reconstructions, and if ignored, they could lead to imaging artifacts. Thus, there is a need for reliable and robust sample preparation techniques for AFM.

The assortment of features of typical AFM substrates currently in use is rather broad. These include silicon surfaces (mica, glass, and silicon wafers) and hydrophobic graphite. Mica is the preferred substrate for AFM. It is a layered mineral in which sheet structures are basic units consisting of two polymerized sheets of silica (SiO_4) tetrahedrons. Two such sheets are juxtaposed with the vertices of their tetrahedrons pointing toward each other; the sheets are cross-linked with cations (e.g., aluminum in muscovite), and hydroxyl pairs complete the coordination of these cations. The negatively charged layer packages $Al_2(OH)_2[AlSi_3O_{10}]$ are held together by K^+ cations (Box 2.2). The adjacent sheets of mica can be split, and K^+ cations are shared between two surfaces, so the split leads to uncompensated charges on separated surfaces, giving the mica surface a negative charge. The major attractive feature of mica is its atomically flat topography. However, there are a few important features of mica as a convenient substrate for AFM: First, mica sheets can be cut with scissors or a custom cutter to make a sheet with desired size and shape. Second, because of the layered structure and strong interaction within the sheet, the cleavage of mica leads to a flat surface that is atomically flat over micron-sized areas. Third, the cleaved surface is free of impurities, so there is no need for additional cleaning of the surface for the surface chemistry. Note that other silicon surfaces such as glass or silicon wafers have to be thoroughly cleaned with strong oxidizers prior to their use as imaging substrates.

Box 2.2 Mica Structure

Micas have sheet structures whose basic units consist of two polymerized sheets of silica (SiO_4) tetrahedrons. Two such sheets are juxtaposed with the vertices of their tetrahedrons pointing toward each other; the sheets are cross-linked with cations (e.g., aluminum in muscovite; blue triangles in Figure 2.13), and hydroxyl pairs complete the coordination of these cations. The negatively charged layer packages $Al_2(OH)_2[AlSi_3O_{10}]$ are held together by K^+ cations (large yellow circles in Figure 2.13). The adjacent sheets of mica can be split, and K^+ cations are shared between two surfaces, so the split leads to uncompensated charges on separated surfaces, giving the mica surface a property of a negatively charged surface. The exposed silicon oxide groups undergo hydrolysis and form a hydroxyl group that interacts with silanes and silatranes. Additional treatment of mica with plasma facilitates the hydrolysis reaction.

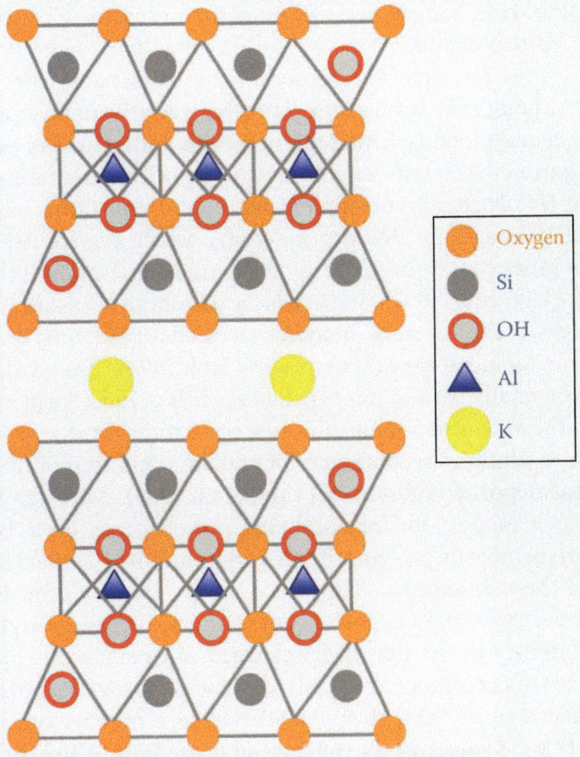

Figure 2.13 Mica structure. Saffron rings, O atoms; grey rings, Si atoms; red rings, OH molecule; blue triangles, Al atoms; and large yellow rings, K atoms.

There are many cases in which unmodified substrates including freshly cleaved mica are used as a major substrate. A few examples describing this are presented in the following sections.

2.5.1 Use of Nonfunctionalized Surfaces

2.5.1.1 AFM of Membranes

Model membranes (Langmuir–Blodgett [LB] films) can be assembled from solutions on different substrates, such as mica, silicon wafers, glass, and gold surfaces. At early stages of biological AFM development, Weisenhorn et al. (1991) originally demonstrated that AFM can image individual lipid molecules if low forces are applied to the scanning tip. They used silylated silicon wafer as a substrate to prepare LB films formed by a β-helical oligopeptide containing alternating D- and L- residues of valine (Weisenhorn et al. 1992). AFM studies revealed that the axis of the oligopeptide helix is perpendicular to the membrane surface.

2.5.1.2 AFM Imaging of Proteins

High-resolution AFM images were obtained for a number of membrane proteins, which, after adsorption to or incorporation into lipid monolayers, form two-dimensional arrays. The large, flat membrane sheets can be attached to an appropriate substrate although the strong interaction between the tightly packed molecules reduces their lateral mobility. Using this approach, initial images were obtained for bacteriorhodopsin, which self-assembles into a two-dimensional crystalline purple membrane of *Halobacterium halobium* (Butt et al. 1990). Other examples of high-resolution AFM images include gap junctions, which are specialized eukaryotic plasma membranes containing an array of channels that connect the cytoplasm of adjacent cells (Hoh et al. 1993). These bilayer membranes adsorb well to glass and mica, and high-resolution AFM imaging revealed a hexagonal array of channels in the gap junction membrane (Hansma and Hoh 1994). Spectacular examples are images of hexagonally packed intermediates (HPIs), which form a standard array of protein on the outer membrane of *Deinococcus radiodurans*.

Molecular resolution was obtained for cholera toxin immobilized in an artificial membrane deposited onto mica (Yang et al. 1993). Although the use of lipid membranes as a support for immobilizing proteins may be helpful for sample preparation of membrane proteins, in general it cannot be applied to water-soluble proteins and their ensembles. The Shao group has produced high-resolution images of the pertussis toxin deposited onto bare mica (Yang et al. 1994). This protein's strong affinity to the negatively charged mica surface is likely due to the fact that the toxin's receptors are positively charged at normal pH. Additionally, it was demonstrated in Yang et al. (1994) that in a number of cases, bare mica can be a rather good substrate for water-soluble proteins. A two-step procedure of mica modification, which allowed the authors to get stable AFM images of water-soluble ferritin molecules, was suggested by Ohnishi et al. (1992).

2.5.2 Surface Functionalization for AFM Studies

Surface functionalization is widely applied in AFM force spectroscopy with various approaches reviewed in Riener et al. (2003), with an approach utilizing self-assembled monolayers formed on a mica surface described in Berquand et al. (2005) also worth noting. Hydrophobic silicon wafer surfaces obtained by treatment with hexamethyldisilazane were used by Ohnishi et al. (1992) for AFM studies of ferritin molecules. Such a substrate was used as a support for binding films of poly(l-benzyl-L-histidine) with attached ferritin molecules that formed at the air/water interface. Imaging of LB films formed by helical pentadecavaline utilized a similar modification of silicon wafers (Weisenhorn et al. 1992). Silicon wafers have been hydrophobized with dichlorodimethylsilane and further used as a substrate for AFM imaging of fibrinogen (Wingren et al. 1991). The coating of glass slides with collagen was critical for the attachment of glial cells (Henderson et al. 1992). This procedure allowed the authors to image dynamics of actin filaments in living cells with AFM. A procedure to functionalize silicone rubber was developed for the covalent immobilization of fibronectin (Volcker et al. 2001). Using a plasma-based surface modification of polymer materials with desirable bulk properties, Vohrer et al. prepared surfaces with tailor-made surface properties (Vohrer et al. 2003).

AFM was very attractive for imaging DNA, RNA, and their complexes with proteins, but the progress in their AFM imaging was made only after development of reliable sample preparation methods that are described later in this chapter. Mica, due to its atomic flatness, was the most attractive substrate, but a freshly cleaved mica surface is negatively charged and cannot bind to negatively charged DNA or RNA molecules. Vesenka, Bustamante, and coworkers (Bustamante et al. 1992; Keller et al. 1992; Vesenka et al. 1992) applied the electron microscopy method of ionic treatment of mica by treating the surface with multivalent cations (e.g., Mg^{2+}) to increase its affinity to DNA, which is held in place, permitting reliable imaging by AFM. Dramatic improvements to this approach were obtained by using specially made tips and by imaging under propanol (reviewed in Hansma and Hoh 1994). Other di- or trivalent cations can be used for pretreatment of mica (Thundat et al. 1992; Hansma and Hoh 1994; Bezanilla et al. 1995). The presence of a multivalent cation helps spread the DNA and bind it to the mica. It was shown later that mica pretreatment could be avoided provided that divalent cations are present in the buffer solution. Other approaches were also developed. The simplicity of the cation-assisted approach led to the progress of imaging various protein–DNA complexes. Bustamante and coworkers obtained AFM images of DNA and DNA-protein complexes with regulatory proteins (i.e., RNA polymerase and chromatin) (e.g., Bustamante and Rivetti 1996; Singh et al. 1996; Bustamante et al. 1997; Rivetti et al. 1998, 1999, 2003; Guthold et al. 1999b). This sample preparation procedure was used to study DNA degradation with nuclease (Bezanilla et al. 1994) and RNA polymerase movement along DNA (Guthold et al. 1994, 1999b). However, the requirement for the presence of divalent cations was a limitation that in a number of cases was a serious obstacle. Chemical treatments of mica were useful alternatives.

One of such approaches utilizes treatment of freshly cleaved mica with 3-aminopropyltriethoxysilane (APTES) to functionalize the mica surface with amine groups (AP-mica) (Lyubchenko et al. 1992a, 1992b, 1993a, 1993b, 2001). Schematically, this functionalized surface is shown in Figure 2.14a. The major advantage of this sample preparation procedure is that it works under a wide variety of ionic conditions, such as varying pH and temperature. The AP-mica technique is expected to be a suitable surface for studying the DNA binding mechanisms with different types of ligands. The characteristics of AP-mica made possible the imaging of nucleic acids (DNA, RNA, and kinetoplast DNA) and various nucleoprotein complexes (reviewed in Lyubchenko et al. 1995, 2001; Lyubchenko 2004). It was found also that, once prepared on AP-mica, samples are stable and do not absorb any contaminants for months while requiring minimal precautions for the sample's storage (Lyubchenko 2004). Images of circular DNA and reconstituted chromatin shown in Figure 2.12 were obtained with the use of an AP-mica procedure. However, a few weaknesses with the use of APTES, and apparently other silanes, such as hydrolysis and polymerization of APTES at neutral pH, were identified and a useful replacement was proposed and developed (Shlyakhtenko et al. 2000b, 2003a). The alternative procedure functionalization is based on the use of 1-(3-aminopropyl)silatrane (APS) instead of APTES. A suggested mechanism of the surface modification is shown in plate B of Figure 2.14, Box 2.3. Unlike APTES, 1-(3-aminopropyl) silatrane (I) is more

Box 2.3 Mica Functionalization with Silanes and Silatranes

A weak cationic surface is obtained if 3-aminopropyltriethoxysilane (APTES) is used to functionalize the mica surface with amino groups (AP-mica). The schematic for the chemical reaction of APTES with mica is shown in Figure 2.14a. As a result, functionalized surfaces remain positively charged at pH below pKa values (pKa = 10.4) and are capable of binding to negatively charged DNA in the pH range of stable DNA duplexes.

An analogous surface chemistry utilizes a more hydrolytically stable silatrane reagent 1-(3-aminopropyl) silatrane (APS) instead of silanes. Slow hydrolysis of the silatrane moiety allows one to avoid clumping and enables a smooth modification of the surface with amino groups achieving the results similar to vapor treatment. The schemes for the synthesis of APS and its reaction with the mica surface are shown in Figure 2.14b. Reaction with the surface is apparently proceeding through several steps (phases 1–3) with eventual loss of triethanolamine molecule and covalent attachment of the 3-aminopropylsiloxane group to mica. Both methods are robust and work reproducibly in various topographic studies involving DNA.

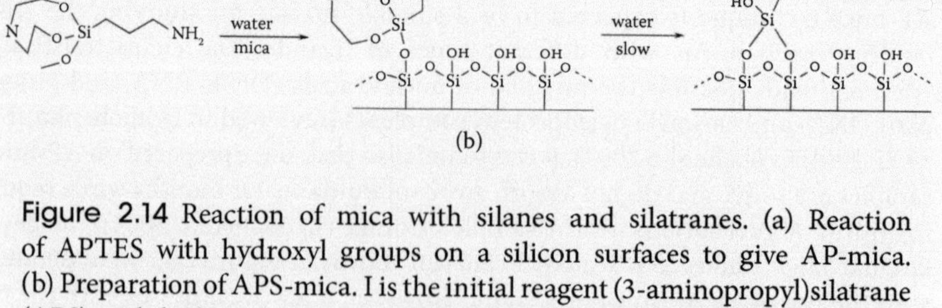

Figure 2.14 Reaction of mica with silanes and silatranes. (a) Reaction of APTES with hydroxyl groups on a silicon surfaces to give AP-mica. (b) Preparation of APS-mica. I is the initial reagent (3-aminopropyl)silatrane (APS) with hydroxyl groups on a silicon surface. The initial adduct II may react with a second surface OH group forming III in a reversible equilibrium.

stable and extremely resistant to hydrolysis and polymerization at neutral pH. The final product (III, APS-mica) is obtained after a partial hydrolysis of the intermediate II on the surface. Apparently, the stabilizing triethanolamine unit remains attached to the Si atom in APS mica since the linkage to the surface is more robust toward hydrolysis than the surface functionalized with APTES. Later sections in this chapter outline a number of projects in which the use of APS-mica was critical in elucidating structural properties of DNA and protein–DNA complexes. The following useful features of AP-mica and APS-mica procedures were identified: (1) high reproducibility of the sample preparation; (2) deposition of the sample in the variety of ionic strengths, pH levels, and temperatures; and (3) no need for divalent cations for deposition.

2.6 AFM Imaging of DNA and Protein Complexes

DNA was the main test sample in adjusting AFM methodology for further visualization studies of complexes of DNA with proteins. Prior to the AFM invention, electron microscopy (EM) was the main tool for such studies; however, relatively complex and harsh sample EM preparation procedures and the lack of being able to visualize molecules at conditions close to physiologic conditions was the major concern with the EM applications. The AFM methodology as described above is significantly gentler. Importantly, the samples are prepared in buffer solutions required for the complex's stability, including different pH and high and low salt conditions. The samples imaged with AFM are scanned at environmental conditions, which is a great advantage when compared with EM which operates in vacuum and requires drying of the sample, making its preservation a major problem with EM visualization. A few examples of AFM applications to characterization of DNA and protein–DNA complexes are presented in the following sections. Demonstration of AFM operation in aqueous solutions opened novel avenues for time-lapse nanoimaging studies not amenable to EM; Section 2.7 is dedicated to such AFM studies.

2.6.1 Visualization of Plectonemic Structure of Negatively Supercoiled DNA

In 1963 Dulbecco and Vogt and Weil and Vinograd made a fundamental discovery. They found that polyoma virus DNA is not a linear but a covalently closed circular molecule. Numerous further studies have led to the conclusion that circular topology rather than linear topology is the predominant DNA form *in vivo*. There are a number of properties of circular DNA topology, and the sensitivity of the global DNA conformation to the local structural changes is one of them. Local unwinding of the DNA duplex is accompanied by overwinding of the entire circle into a more compact shape termed *supercoiled DNA*. If the DNA duplex overwinds by n number of turns, this duplex rotation is compensated by interwinding of circular DNA in the opposite direction by the formation of superhelical turns. Importantly, circular plasmid DNA in cells is negatively supercoiled, and this DNA state is tightly linked with various DNA functions. We mention just one; many genetic events require communication between proteins bound to distant sites on DNA including DNA replication, control of gene expression through loop formation, site-specific recombination, and transposition. A fundamental feature of supercoiled DNA geometry is that distantly separated DNA regions can be brought into close proximity (Shlyakhtenko et al. 2000a; Vologodskii 2015). Obviously, direct visualization of such global DNA conformations is important for understanding the biological function of DNA supercoiling. Initially, EM was the only visualization method capable of visualizing supercoiled DNA, but inflexibility of the sample preparation method was a limiting factor in studies focused on labile features such as DNA supercoiling, because the shape of the molecule was very sensitive to such factors as ionic strength and pH. These are not issues for AFM; therefore, this technique was successfully applied to the imaging of supercoiled DNA (Lyubchenko and Shlyakhtenko 1997; Shlyakhtenko et al. 2000a, 2003b; Potaman et al. 2002; Lyubchenko 2004, 2011; Mikheikin et al. 2006; Lyubchenko and Shlyakhtenko 2009).

Figure 2.15 shows images of plasmid DNA, 5.6 kBp long, obtained with AFM operating in tapping mode. Frame a shows three different DNA molecules, and the higher-resolution image of DNA is shown in frame b, top-left. In this image the DNA strands are interwound, forming the plectonemic morphology of supercoiled DNA. The loops between the intersections of the strands are of relatively similar sizes, and such images are in line with computational models of supercoiled DNA (Higgins and Vologodskii 2015). Two other molecules in frame a show an interesting branched morphology of the supercoiled DNA, which is expected for molecules of such sizes (Lyubchenko 2011). DNA is a negatively charged polymer with a rather high linear charge density, and electrostatic repulsion of the strands plays a critical role in the overall geometry of DNA molecules. This electrostatic effect can be modulated with counterions; therefore, one

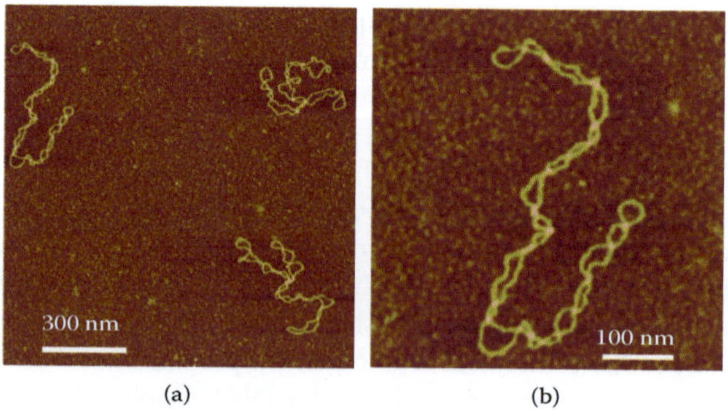

Figure 2.15 AFM images of supercoiled DNA (supercoiling density s = −0.075). (a) A large-scale image and (b) high-resolution image obtained by rescanning around the molecule on the left. The samples after the deposition onto AP-mica were rinsed with water and argon-dried. The images were acquired in air with a NanoScope IIImicroscope operated in Tapping Mode. (Reproduced from *Micron*, 42, Lyubchenko, Y. L., Preparation of DNA and nucleoprotein samples for AFM imaging, 196–206, 2011, Copyright (c) (2010), with permission from Elsevier.)

should expect the effect of ionic strength on the geometry of supercoiled DNA. AFM studies directly confirmed this assumption. It was shown by Shlyakhtenko et al. (2003b) that at high ionic strength, the intersegmental distance in the supercoiling loops decreases, forming very close DNA-DNA contacts. As mentioned previously, DNA supercoiling is a mechanism by which distant DNA segments can be brought in close proximity to one another, and such juxtaposition is important for numerous genetic processes such as site-specific recombination. Importantly, the formation of tight contacts in the AFM experiments was observed at cation concentrations close to those found physiologically.

Worth noting is that AFM is the only imaging technique that has directly demonstrated the dependence of the geometry of supercoiled DNA on ionic strength. However, a careful selection of the sample preparation is the key in these studies. The results described previously were obtained with the use of an AP-mica procedure. An alternative cation-assisted technique can lead to artefacts. Note the study by Nagami et al. (2002) in which AFM images of supercoiled plasmid DNA were obtained with the use of Mg^{2+}-treated mica. On images obtained in that publication, supercoiled DNA molecules appear with long, tightly twisted segments, which are interspersed with large loops indicated with arrows and arrowheads. This is in contrast with the data in Figure 2.15, and such morphology of DNA is not supported by numerous studies including EM imaging. These studies suggest that the interaction of DNA with the surface in the cation-assisted sample preparation method changes the morphology of supercoiled DNA. Given the fact that negatively supercoiled DNA is a natural state of the DNA inside cells, studies of such molecules with the use of Mg-assisted technique can lead to artefacts, and this issue should be taken into account when studying DNA samples with AFM. Moreover, additional features of the Mg^{2+}-treated mica must be noted. The data in the study by Rivetti et al. (1996) showed that the persistence length of the DNA deposited with the use of Mg^{2+} method is larger than the persistence length of DNA in solution. Such straightening of DNA during the interaction with the substrate can contribute to the rearrangement of the strands of plasmid that may lead to changes of labile morphology of supercoiled DNA.

2.6.2 Visualization of Alternative DNA Structures in Negatively Supercoiled DNA

Formally, the effects of supercoiling on the structure of DNA can be divided into two main categories: local and global. The global conformational changes of supercoiled DNA were described earlier in this chapter; here we describe AFM studies of local changes to DNA conformation that refer to the transition of a short DNA region from a canonical B conformation into a different one, the "alternative structures," such as left-handed Z-DNA, cruciforms, intramolecular triple helices, and unwound regions.

2.6.2.1 Imaging of Cruciforms

A cruciform is formed in DNA sequences with twofold inverted repeat or palindromic symmetry under unwinding torsional stress from negative DNA supercoiling (reviewed in Sinden 1994). Cruciforms are critically involved in various genetic processes such as initiation of DNA replication (Pearson et al. 1996). The putative equivalence of cruciforms and the Holliday junction, an intermediate in homologous DNA recombination, are another motivation for structural studies of cruciforms (Shlyakhtenko et al. 1998). Understanding the characteristics of cruciforms and their impact on global DNA structure and dynamics may help to explain their biological functions.

There has been little attempt to analyze the spatial geometry of cruciforms. Cruciforms are not thermodynamically stable in linear DNA due to branch migration. Electron microscopy has been used to visualize cruciform structures, but this system involved totally palindromic supercoiled circular DNA, in which the size of cruciforms varied considerably. AFM filled this gap through successful visualization of cruciforms as was done by Shlyakhtenko et al. (1998).

Plasmid pUC8F14C DNA containing a 106 bp F14C inverted repeat with an expected length of the cruciform arms of 53 bp was analyzed by AFM. Figure 2.12a shows one of the AFM images of the sample prepared by deposition from a low salt solution (TE buffer). The main feature of these DNA images is the presence of rather long extrusions (hairpin arms) indicated by arrows. Note that these extrusions have not been observed in supercoiled DNA without the F14C inverted repeat. Several additional facts suggest that these extrusions are the cruciforms of the 106 bp palindrome. First, the size of the arms for extended extrusions is 15–20 nm, in agreement with the expected length of the hairpins containing 53 bp (18 nm for B-helix DNA geometry). Second, extrusions have been unambiguously identified in more than 80% of the DNA molecules, in agreement with two-dimensional (2D) gel electrophoresis data. Third, the number of molecules with extrusions is less for a sample with lower superhelix density. Significantly, two classes of cruciforms were observed. Cruciforms in an extended conformation (Figure 2.12a; see also molecule 2 in Figure 2.16a) are characterized by an angle between the hairpin arms of ca. 180°. A second class, which represented 70%–80% of the population, contained molecules in which the arms form a rather acute angle (X-type geometry, molecules 1 and 3 in Figure 2.16a), with the main DNA strands being sharply bent. The experiments were performed with the use of an AP-mica procedure permitting the deposition and imaging of the sample in a broad range of environmental conditions. Specifically, it was shown that in the presence of high salt or Mg cations, a compact, X-type conformation is highly preferable to its extended counterconformation. Statistical analysis of the interarm angle led to the conclusion on the high mobility of the cruciform arms. This conclusion was confirmed by imaging in an aqueous environment, allowing for the direct observation of cruciform dynamics. These data showed that the X-type conformation was very mobile, allowing the hairpin arms to move between orthogonal and almost parallel orientations.

2.6.2.2 Cruciform as a Molecular Switch for Global Dynamics of Supercoiled DNA

As we described above, the fundamental properties of circular DNA are cross-talk between local DNA conformations and the global structure of the entire molecule. AFM studies of this cross-talk (Shlyakhtenko et al. 2000a) revealed

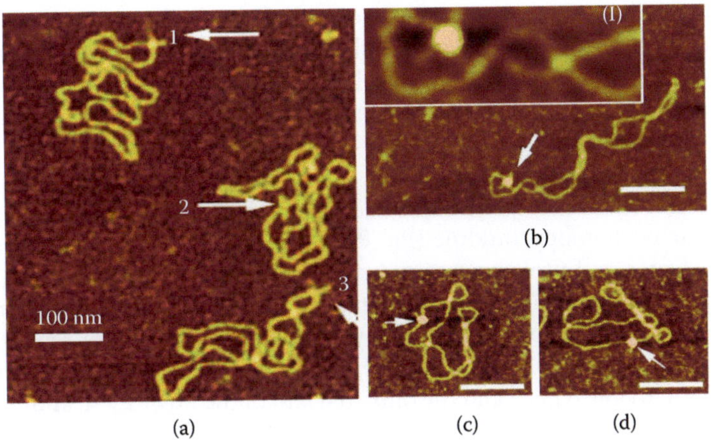

Figure 2.16 AFM images of cruciforms in their complexes with RuvA protein. (a) pUC8F14C plasmid DNA. Cruciforms are indicated with arrows and numbered. (b–d) pUC8F14C plasmid DNA with RuvA protein. Cruciforms in complexes with the protein are indicated with arrows. The scale bar is 100 nm. (Reproduced from *J. Mol. Biol.*, 296, Shlyakhtenko, L.S., et al., A cruciform structural transition provides a molecular switch for chromosome structure and dynamics, 1169–1173, 2000a, Copyright © (2000), with permission from Elsevier.)

a new mechanism of cruciforms, termed *molecular switch* by the authors. AFM images in Figure 2.16a (see molecules 1 and 3) show that the X-type cruciforms tend to be localized at the loops of the plectonemic superhelix due to a bend in supercoiled DNA. Therefore, the X-type of cruciform uniquely orients the plectonemic superhelix in such a way that it occupies an apical position of the molecule defining the overall spatial geometry of a DNA superhelix. This mechanism was tested by Shlyakhtenko et al. (2000a) by AFM imaging of cruciform conformation that was controlled through the addition of RuvA protein, which stabilizes the unfolded conformation of the cruciform. The most striking feature of the RuvA-unfolded cruciforms is their localization in the molecule. As shown in Figure 2.16b through d, unlike X-type cruciforms, unfolded cruciforms stabilized by RuvA exhibit no preference for an apical position of the superhelix. Statistical analysis revealed that without RuvA bound, 94% of DNA molecules contain cruciforms at an apical position, compared to only 6.5% in the RuvA-bound cruciforms. These data suggest that the reptational motion of the DNA chains in supercoiled DNA is restricted by a cruciform in a compact X-type geometry, but this limitation in DNA motion is eliminated as soon as the cruciform adopts an extended conformation.

As mentioned above, DNA supercoiling facilitates a juxtaposition of two distant sites along a molecule needed for DNA transactions such as site-specific recombination, DNA transposition, and gene regulation through loop formation. Contrary to a sequence-specific curved segment that induces a permanent bend, cruciforms exist in the molecules transiently, so the sequence-dependent orientation of the superhelix can be turned on or off depending on cruciform formation. Thus, cruciforms may play the role of a molecular switch for DNA transactions.

2.6.2.3 Alternative Structures in Supercoiled DNA Visualized with AFM

Homopurine–homopyrimidine (Pu–Py) tracts are overrepresented in genomes of eukaryotes and prokaryotes, and several classes of Pu–Py binding proteins have been described (reviewed in Sinden 1994). DNA supercoiling induces a transition in the homopurine–homopyrimidine (Pu–Py) mirror repeat tracts to an intramolecular triple helix (triplex), also termed *H-DNA*. The major elements of H-DNA are the triplex and the unpaired half of the Pu or Py strand. A number of models explaining a role for intramolecular triplexes in gene expression, DNA replication, recombination, and stabilization of the structure of telomeres (reviewed in Mirkin 2008) have been suggested. Prior to the AFM studies in Tiner et al. (2001), the structural visualizations of intramolecular triplexes within plasmid DNA were not possible. Acidic pH values stabilize intramolecular triplexes; therefore, plasmid samples were prepared at acidic pH, and control samples were prepared at neutral pH. Both samples were deposited onto APS-mica. The AFM image for the sample is shown in Figure 2.17a. The formation of a clear kink with a short protrusion indicated with arrows is a distinct feature of the sample prepared at an acidic pH. The formation of a sharp kink is fully consistent with the model of intramolecular DNA triplex. Note as well an apical location of H-DNA structure suggesting that H-DNA formation, similarly to the cruciforms, can also play a role of conformational switch of supercoiled DNA. AFM studies were also performed with a long, imperfect pyrimidine–purine (Py–Pu) insert as described

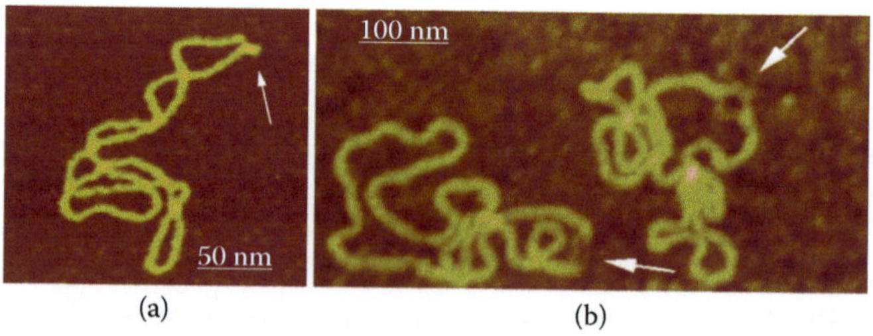

Figure 2.17 AFM images of various alternative DNA structures. (a) AFM images of H-DNA stabilized by negative DNA supercoiling. The H-DNA structure is indicated with an arrow. (b) AFM images of bulges in supercoiled DNA formed by the supercoil induced dissociation of the duplex at the $(ATTCT)_{29}$- region. Open regions are indicated with arrows. (Reproduced from *Biochemistry*, 41, Potaman, V.N., et al., Site-specific labeling of supercoiled DNA at the a+t rich sequences, 13198–13206, 2002, Copyright © (2002), with permission from Elsevier.)

in Kato et al. (2002). These studies made evident that this insert adopts H-DNA conformation, although spatial morphologies varied.

Negative DNA supercoiling stabilizes such a state of DNA as a local duplex unwinding, which would be of very low probability in linear DNA molecules (Amirikyan et al. 1981). The importance of understanding the mechanism of DNA unwinding is supported by the fact that DNA replication and transcription are biologically important processes. The latter requires unwinding of DNA segments as large as ~20 bp (Zaychikov et al. 1995). AFM visualization of opened DNA segments was reported in Potaman et al. (2002, 2003). The repeat $(ATTCT)_{29}$ was cloned into the plasmid, and the plasmid was imaged with AFM under conditions facilitating the local duplex opening. An example of such alternative DNA structures is shown in Figure 2.17b. Opened regions are indicated with arrows. Remarkably, this local DNA structure is dramatically different from H-DNA and the cruciforms. A strong kink, a characteristic feature of H-DNA, is not present in bulged DNA. A stable existence of unpaired regions in DNA at the torsional stress corresponding to the DNA supercoiling *in vivo* prompted Potaman et al. (2002, 2003) to suggest that unpaired DNA regions are critically involved in the initiation of DNA replication. The sequence studied in that paper was a pentanucleotide repeat expansion associated with spinocerebellar ataxia disease. The finding on the stable existence of such repeats in DNA led to the model of the disease-prone expansion of ATTCT pentanucleotide motif (Potaman et al. 2002, 2003).

2.6.3 AFM Studies of Protein–DNA Complexes

Since proven reliable in the imaging of DNA, AFM has also been applied to the study various protein–DNA complexes. In the following sections, we illustrate AFM application with two examples that led to new biology-related discoveries.

2.6.3.1 Structure of Synaptic SfiI Protein–DNA Complex

SfiI is a site-specific DNA binding protein that binds two DNA recognition sites to accomplish the cleavage and can play the role of a model system for studies of site-specific DNA recombination systems (Lushnikov et al. 2006). In the majority of cases, SfiI binds to two distant sites in the same DNA molecule, which are separated by some distance (*cis* complexes); this complex formation should lead to the looping out of the DNA segment between the two recognition sites, and AFM images in Figure 2.18a illustrate the formation of such a *cis* complex. DNA cleavage requires Mg^{2+} cations, but in the presence of Ca^{2+} cations restriction is terminated, although the complex is formed. Therefore, the assembly of synaptic complex was studied in the presence of Ca^{2+} cations. A zoomed image of one of such complexes is shown in Figure 2.18b. The protein appears as a bright globular feature holding two DNA sites. The length measurements were performed to validate a specific binding of SfiI to such a DNA substrate. Additionally, the protein stoichiometry was determined. This was possible due to the ability of the AFM to measure the molecules in 3D, which allowed for the molecule size to be determined from volume measurements of the protein. However, the AFM due to the tip convolution effect requires the calibration of the volume values against protein molecules of known molecular weight to evaluate the protein mass.

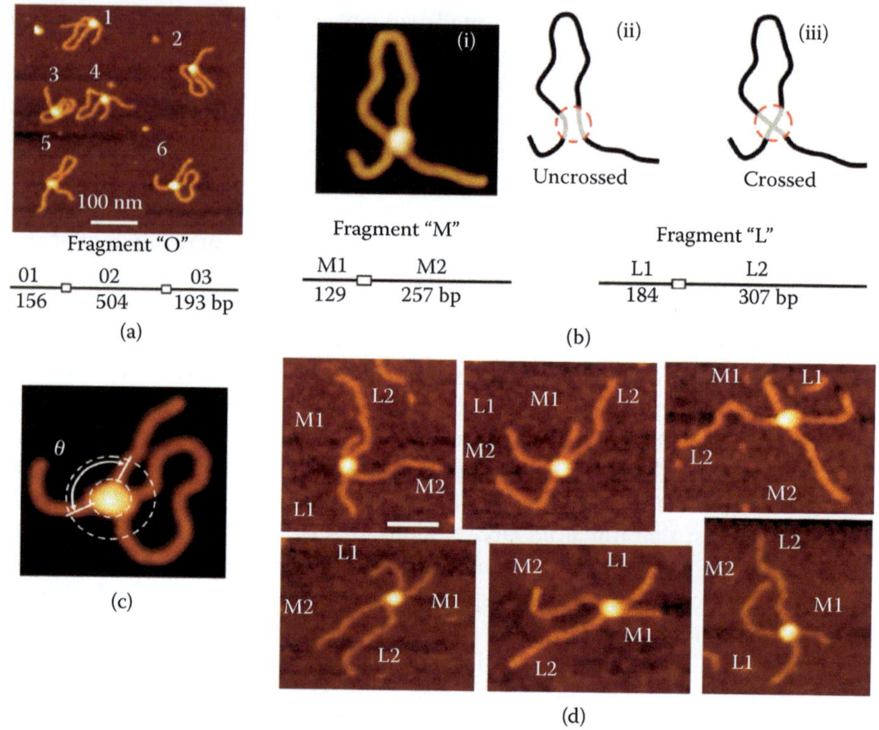

Figure 2.18 AFM study of SfiI-DNA complexes. (a and c) AFM images of synaptic SfiI-DNA complexes formed between two recognition sites separated by 504 bp (left). (b) Models of the possible arrangement of DNA strands in the synaptic complex SfiI-DNA. (d) Complexes formed by SfiI by synapsis of M and L DNA fragments (*trans*-type complexes). The schematic for the fragments is shown on the top, and the AFM images are below. The flanks M1, M2 and L1, L2 are assigned according to their length measurements. (Reproduced with permission from Lushnikov, A.Y., et al., *Biochemistry*, 45, 152–158, 2006, Copyright © (2006) American Chemical Society.)

The volume measurements showed that SfiI binds to DNA as a tetramer; a finding that was in line with the established biochemical data.

One of the questions of great importance for site-specific DNA recombination systems is how the DNA recognition sites are arranged within the protein. Two orientation modes of the binding sites, side-by-side and crossed, are possible, and they are schematically shown in the middle of Figure 2.18c. The left model shows side-by-side assembly; the right panel shows the crossed orientation of the recognition sites. The AFM image shown to the right of the scheme (Figure 2.18c) does not allow one to distinguish between the two binding modes. The authors developed an approach that enabled one to identify such a binding model. The approach is illustrated in Figure 2.18d. The synaptic complex is formed in *trans* mode in which two DNA substrates M and L containing only one specific site for

SfiI are used. However, with the use of two substrates, homologous complexes (M-M and L-L) along with heterologous M-L complexes are formed. The complexes can be identified by the length measurements if fragments L and M differ in the lengths and the sizes of flanks between the specific site and the ends of the fragments. Schematics for M and L fragments designed with the arm lengths as shown in the top of the figure made it possible to differentiate each complex by the length measurements. AFM images graphically illustrate how the assignments can be made. Both DNA substrates were mixed together and along with SfII, *trans*-type intramolecular complexes were assembled. The AFM images in Figure 2.18d show a few typical examples of such *trans*-complexes. Statistical analysis of the length measurements over hundreds of such complexes led to the conclusion that the DNA strands in the SfiI synaptosome are crossed.

AFM images for *cis* and *trans* complexes (Figure 2.18a and d, respectively) show that DNA arms are oriented at some angle, thereby suggesting that the recognition DNA sites within the synaptosome are oriented at an angle. The angle measurements demonstrated that the mean angle values are grouped between 60° and 120°. This finding led the authors to two conclusions. First, the DNA segments are not parallel, but oriented at a specific angle. Second, the double peak distribution suggests that two orientations of the DNA recognition site are possible. Both populations appeared to be equally represented, suggesting that the two possible orientations of the helices were formed almost equally or, in other words, that there is no preferential orientation of the DNA cognate site within the complex. The entire DNA binding region (5′-GGCCNNNNNGGCC-3′) is not fully symmetric, so the AFM data on the symmetry of SfiI binding suggest that the central part of the recognition site on DNA is not essential for binding specificity and the protein very likely due to formation has no direct contact with the bases inside the NNNNN motif of the DNA spacer. This AFM prediction and the arrangement of the DNA segment in complexes with SfiI were perfectly in line with the crystallographic model for SfiI-DNA complexes, which appeared independently of the AFM studies (Vanamee et al. 2005).

It is important to note that the Mg-assisted technique does not work here, as the enzyme will cleave DNA in the presence of Mg^{2+} cations. However, it was not a problem for the APS-mica procedure, so the synaptic complexes were assembled and analyzed in the presence of Ca^{2+} cations to prevent DNA cleavage.

2.6.3.2 Role of SSB Protein in Targeting of Stalled Replication Fork by RecG Helicase

DNA replication is a complex process in which the movement of the replication fork by the replication machinery is interspersed by pauses during which problems with the DNA template are repaired (Sun et al. 2015). The major repair mechanism includes the reversal of the replication fork and formation of the Holliday junction, a target for the homologous DNA repair machinery, allowing it to repair the defect leading to the replication stall. ATP-dependent RecG helicase is the key player of the reversal fork movement, and this protein has a high affinity to the stalled replication fork. Crystallographic data for RecG in complex with fork DNA led to a model for the fork recognition in which one of the RecG domains interacts with the replication fork, while two other domains bind to the parental DNA duplex (Singleton et al. 2001). However, these data do not answer a number of questions. The RecG recruitment is one of them. There are a few proteins per cell, so one can ask the question of how RecG recognizes a stalled fork efficiently. Second, it was shown that the activity of RecG in the early stages of fork rescue is enhanced and controlled by the single-stranded DNA binding protein (SSB) (Bianco 2015), but the mechanism of this interaction remains unclear. These questions were addressed by Sun et al. (2015) in their study where the stalled DNA substrates were designed, and the interaction of the substrates with SSB and RecG was characterized by AFM.

The DNA substrate contained a 3′-end 69 nt single-stranded DNA (ssDNA) segment inserted between two DNA duplexes of different lengths (255 and 355 bp). According to the crystallographic model, two domains of RecG form a complex with the parental duplex DNA ahead of the fork, while the wedge domain interacts with the ssDNA segment of the fork. The substrate binds SSB protein as shown in the AFM image (Figure 2.19a) in which SSB appears as bright

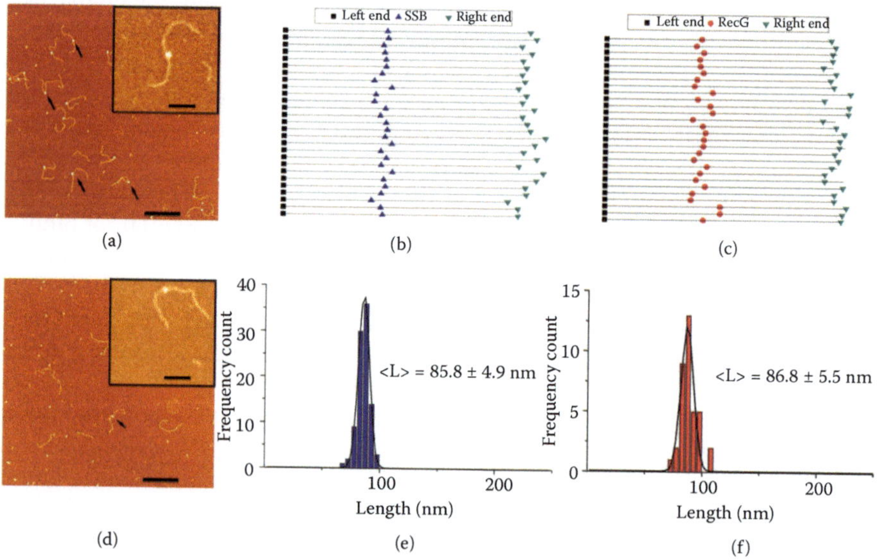

Figure 2.19 AFM analysis of the interaction of SSB and RecG with the fork DNA substrate. (a and d) Representative AFM images of SSB and RecG only on the fork DNA. Arrows point to the complexes. Bar size is 200 nm. Insets show enlarged images of the complexes. Bar size = 50 nm. (b and c) Maps for the positions of SSB and RecG, respectively, on the DNA substrate. The DNA molecules were aligned by the left end of the parental flank without normalization of the molecule lengths. (e and f) Distributions of the proteins' distances from the left end for SSB ($n = 95$) and RecG ($n = 37$), respectively. (Reproduced from Sun, Z., et al., *Sci. Rep.*, 5, 9625, 2015, an open access publication: permission has been distributed under a Creative Commons CC-BY license. With permission.)

feature located within the DNA substrate. The yield of complexes was as high as 85.3 ± 3.8%. The locations of SSB on the DNA molecules were mapped, and the results in Figure 2.20b and c directly illustrate the correlated position of the molecules or SSB on different molecules corresponding to the fork's expected position within the DNA substrate.

Similar experiments were performed with RecG. In the AFM image shown in Figure 2.19d, molecules of RecG on the DNA can also be seen (indicated by arrows). Even though this is the preferred substrate for RecG, the yield of protein–DNA complexes was only 10%. This is significantly lower than SSB, although a twofold higher concentration of RecG was used. RecG's DNA binding sites were mapped (Figure 2.19d and f). The mean value for the lengths of the left flank of the DNA molecules was 86.8 ± 5.5 nm, corresponding to the position of the fork in the DNA substrate (86.7 nm). A visual comparison of the images for fork DNA-protein complexes shows that the brightness of the SSB molecules is higher than that of the RecG molecules, which is in line with the height and volume measurements for the two proteins. These differences in height and volume become critical in distinguishing SSB from RecG when bound to the same DNA molecule.

Figure 2.20a shows AFM images obtained in experiments in which both SSB and RecG were bound to the same DNA substrate. The new feature in this image is the appearance of complexes containing two molecules, as indicated by arrows. Four zoomed images of double-molecule complexes are shown in plates B–F, in which two molecules with different sizes are indicated with green and blue arrows. The volumes and heights of the molecules in the double-molecule complexes were very close to the values obtained for the individual SSB-DNA and RecG-DNA complexes. These data demonstrate that RecG is capable of binding to the DNA fork in the presence of SSB, because the two proteins appear on the

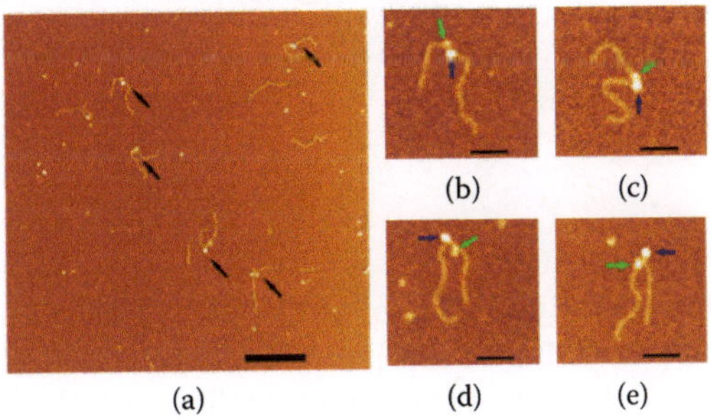

Figure 2.20 AFM images of the complexes made by SSB and RecG with the fork DNA substrate. (a) Large-scale AFM images, in which double-particle features are indicated with arrows. Bar size is 200 nm. Zoomed images (b–e; bar size = 50 nm) of four double-particle complexes. Large and small particles are indicated with blue and green arrows, respectively. The black and green arrows point to SSB and RecG proteins in the double-particle complexes. (Reproduced from Sun, Z., et al., *Sci. Rep.*, 5, 9625, 2015, an open access publication: permission has been distributed under a Creative Commons CC-BY license. With permission.)

same DNA substrate simultaneously. Two additional features emerged from the AFM studies: First, the efficiency of RecG binding in the presence of SSB increases threefold (~30.3% in the double complexes versus 9.7% of RecG-DNA complexes only). Second, RecG in the presence of SSB does not occupy the fork position; rather, statistical analysis of the length measurements showed that the position of RecG is shifted away from the fork position by the mean distance ~35 bp. The translocation activity was not observed when the RecG-DNA complexes were assembled without SSB, suggesting that interaction with SSB is required to allow RecG to translocate along the duplex DNA.

The authors hypothesize the interaction of SSB with RecG induces remodeling of RecG, enabling the protein to translocate along the DNA duplex. According to Figure 2.21a, RecG initially binds the fork using its three DNA binding domains. Then the RecG wedge domain 3 disengages from the fork, but domains 1 and 2 remain bound to the DNA duplex (Figure 2.21b[i]). As a result, RecG loses the "hook" that kept it bound to the fork, allowing the protein to translocate along the duplex (Figure 2.21b[ii]). The mean migration range is as

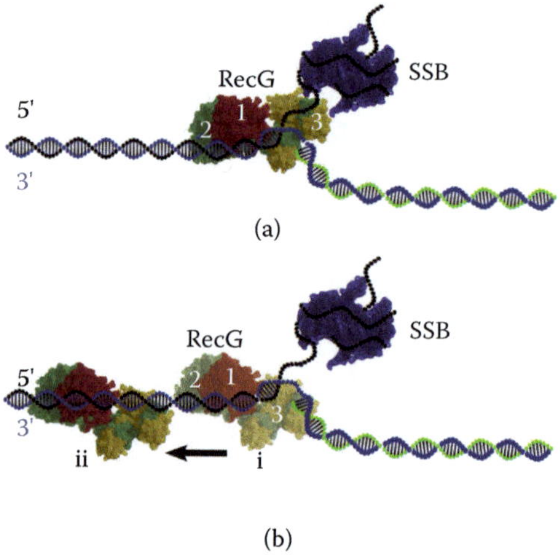

Figure 2.21 Models for RecG binding to the DNA fork in the presence of SSB. (a) No interaction with SSB. In the type 1 complex, RecG binds specifically to all arms of the fork, with domains 1 and 2 binding the parental DNA duplex, whereas the wedge domain 3 binds the single-stranded arm of the fork. SSB occupies the ssDNA arm of the fork. (b) In the type 2 complex formed after SSB remodeling, the wedge domain 3 dissociates from the ssDNA arm, but domains 1 and 2 remain bound to the parental DNA flank, enabling RecG to translocate along the duplex as shown in (i) and (ii). Model (i) shows the RecG position at the fork just after remodeling, and (ii) shows the RecG position after translocation away from the fork. (Reproduced from Sun, Z., et al., *Sci. Rep.*, 5, 9625, 2015, an open access publication: permission has been distributed under a Creative Commons CC-BY license. With permission.)

large as ~35 bp, suggesting that RecG in this binding mode is capable of scanning a rather large DNA segment. Importantly, RecG remains in the fork vicinity and can be recruited for the fork regression when the fork stalls. Apparently during remodeling, RecG switches off its helicase activity that works in the direction opposite to the fork progression and is the mechanism by which RecG remains in the proximity of the replication fork. Note that the fork regression requires ATP hydrolysis (McGlynn and Lloyd 2002), whereas translocation by the remodeled RecG is a thermally driven process and does not require ATP. Therefore, RecG can be recruited rapidly to accomplish its fork regression role as soon as the fork stalls.

Although SSB is considered to be an architectural single-stranded DNA binding protein with no sequence specificity for binding DNA, evidence accumulated recently points to the interaction of SSB with more than a dozen proteins. The experiments described above demonstrate a novel role of SSB in which SSB dramatically increases RecG loading efficiency onto the DNA fork. Moreover, SSB plays a role of a remodeling factor of RecG after which RecG disengages from the fork but remains bound to the parental duplex. Importantly, C-terminus of SSB plays a critical role for both new SSB activities; a finding that falls well in line with early findings on the role of C-terminus with SSB interaction with other proteins. Interestingly, that SSB and RecG do not form stable complexes could be detected by AFM, suggesting that the interaction is transient. Given the fact that SSB is involved in other genetic processes, it is reasonable to assume an active role of SSB is not to be limited to the DNA replication process, but can be found in other processes requiring SSB.

2.6.3.3 AFM to Study Mismatch Repair Apparatus

The DNA mismatch repair (MMR) system is biologically one of the most important systems whose function is to correct the DNA synthesis errors that occur during replication (Modrich 2006). As a result, the fidelity of DNA synthesis increases by several orders of magnitude. Progress in the structure-function study of MMR was signified by the 2015 Nobel Prize to P. Modrich for the MMR discovery and its thorough studies. Various techniques were applied to characterize the MMR system with AFM among them.

In *Escherichia coli*, MutS and MutL are the protein complexes that recognize the lesion, and MutH is recruited to initiate the process by incision of the DNA duplex at the lesion. Erie and colleagues in their early AFM study asked the question on how the MutS recognizes the mismatch (Wang et al. 2003). They applied AFM topographic imaging to identify complexes of MutS with specific complexes using the DNA substrates in which the mismatch position was introduced to a desired location. One of the effects induced by MutS binding was bending of the DNA duplex, and AFM was used to measure the bend angle. They found that MutS binding to a homologous region induces a bend of a defined angle, but binding to the mismatch leads to bent and straight complexes. Based on these studies as well as other biochemical and crystallographic data, the authors proposed a model in which the protein conformation depends on the DNA target type. MutS bends DNA while searching the target and changes the conformation when it hits the target. The protein undergoes a conformational transition leading to the formation of the kinked complex. After that, additional conformational transition occurs after which the DNA in the complex becomes straight. Using a set of MutS mutants, this group later identified motifs responsible for DNA bending and straightening to show that Phe is a key residue in the mismatch recognition pattern (Tessmer et al. 2008). This group also applied AFM to answer questions regarding the low affinity of MutS to the mismatch identified in gel-shift experiments (Yang et al. 2005). They developed the AFM-based method for evaluation of protein–DNA binding constants and applied this methodology to the MutS-DNA complexes. They showed that MutS has a high affinity to the ends of DNA duplexes, and this binding property masked the high affinity of MutS to the mismatch. AFM imaging allowed the authors to identify specific complexes only. The developed approach for measuring the binding constant can be useful for AFM practitioners working in protein–DNA studies.

Recently, another group applied AFM to the characterization of interaction with a mismatch of the complete MutS-MutL-MutH system (Josephs et al. 2015). From the high-resolution AFM images, they were able to identify assembly and disassembly of the repair system. Interestingly, they found that MutS plays a role of the scaffold to which MutL and MutH assemble. The authors also observed bridging of the mismatch and hemi-methylated GATC sites by the Mut-complex binding due to the formation of looped DNA complexes. The ability to observe such a complex and dynamic system with AFM is important for further studies of this and other complex protein–DNA systems.

It is important to illustrate the application of AFM to the characterization of eukaryotic MMR systems. Sacho and colleagues (2008) characterized with AFM a full-size MutLα protein that is a homolog of the prokaryotic MutL protein. This protein contains a long unstructured segment (UNS), making it problematic to study a full-size MutLα with traditional X-ray crystallography. MutLα assembles as dimers, and AFM images in Figure 2.22a illustrate a morphological heterogeneity of the protein. Moreover, the morphology depends on the ADP that causes protein collapse. Frames b through e show a set of conformations. Frame b shows the protein morphology in which three globular features connected with two fibrillar segments that are unstructured segments. More of such three-blob assemblies are shown in Figure 2.22f. The schematic in frame b explains the assembly. It is proposed based on the known properties of MutLα that two monomers interact by their C-terminal globular segments to make that three-blob structure. The structure can collapse to make the assembly shown in frame e through the states shown in frames c and d. The analysis of the AFM data led to the conclusion that during the ATPase cycle in which sequential nucleotide binding, hydrolysis, and release occurs, MutLα undergoes a series of conformational changes. It is hypothesized that conformational changes, modulated by adenine nucleotides, mediate the interactions of MutLα with other proteins in the MMR pathway, coordinating the recognition of DNA mismatches by MutSα and the activation of MutLα with the downstream events that lead to repair.

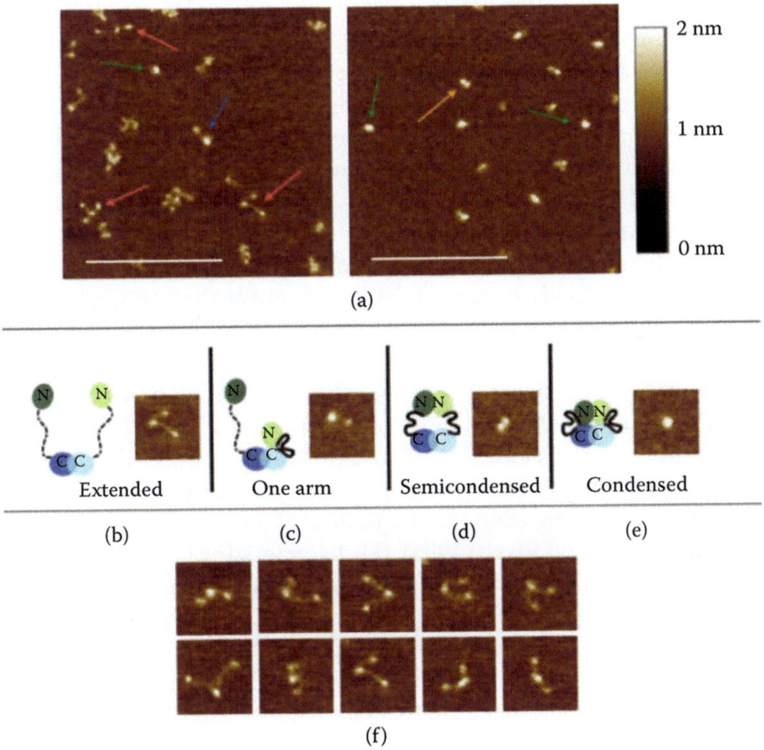

Figure 2.22 AFM images of yMutLα. (a) 500 nm × 500 nm images of MutLα deposited in the absence of nucleotide (left) and in the presence of 5 mM ADP (right). Arrows point to examples of the four different conformational states seen in the images: red = extended; blue = one armed; orange = semicondensed; green = condensed. Scale bar (white) is 250 nm. (b–e) 100 nm × 100 nm images of the four states accompanied by cartoons of MutLα in the different conformational states. In the cartoons, domains are indicated by ovals, connected by a flexible linker. A disordered linker is shown as a dashed line; an ordered linker is shown as a solid heavy line. Mlh1 is shown in light green and blue, and Pms2 (yPms1) is shown in dark green and blue. (f) A selection of 100 nm × 100 nm images of the extended state to show the flexibility of the linker arms. (Reproduced from *Mol. Cell.*, 29, Sacho, E.J., et al., Direct visualization of asymmetric adenine-nucleotide-induced conformational changes in MutL alpha, 112–121, 2008, Copyright © (2008), with permission from Elsevier Inc.)

2.7 Time-Lapse AFM Operation

One of the most attractive features of the AFM for biological applications is the ability of the instrument to acquire images in aqueous solutions. First, the sample is imaged in fully hydrated state, which is extremely important for the imaging of biological samples. Second, dynamics of the sample can be observed from a set of consecutive images. This mode is termed *time-lapse imaging*, and given nanometer range AFM resolution, dynamics of molecules or interaction of the molecules can be visualized. This is a unique property of AFM among all nanoimaging techniques. Finally, imaging in liquid eliminates effects of capillary forces, the major factor complicating imaging at gentle conditions and decreasing spatial resolution of AFM as discussed in Section 2.3. Reliable visualization of molecules with AFM requires their stable binding to the surface. At the same time, the imaging of dynamics at the single molecular level requires loose DNA binding, permitting the molecule to move relatively freely at the surface. Therefore, the sample preparation procedure should be adjusted to reconcile these conflicting requirements.

The history of time-lapse AFM imaging begins with the pioneering work of the Hansma group in which direct observation of fibrin self-assembly in aggregates was performed (Drake et al. 1989). Since this early work, single molecule AFM imaging has been applied to various problems such as (1) observing how RNA polymerase moves along DNA (Bustamante et al. 1994, 1997; Guthold et al. 1994, 1999b; Bustamante and Rivetti 1996); (2) characterizing local and global dynamics of supercoiled DNA (Lyubchenko and Shlyakhtenko 1997; Shlyakhtenko et al. 2003a; Lyubchenko 2004; Mikheikin et al. 2006); (3) directly visualizing Holliday junction branch migration (Lushnikov et al. 2003; Lyubchenko 2004); (4) determining chromatin dynamics (Lyubchenko et al. 2009; Shlyakhtenko et al. 2009b); (5) characterizing conformational dynamics of membrane proteins (reviewed in Muller and Engel 2007; Muller 2008; Scheuring and Dufrene 2010); and (6) performing single molecule detection of receptor dynamics in the membranes of living cells (Heinisch et al. 2010; Scheuring and Dufrene 2010). Novel properties of molecular systems were discovered with these time-lapse studies. These experiments at the same time highlighted a set of problems associated with the time-lapse methodology. A few examples are reviewed in the following sections.

2.7.1 Visualization of Holliday Junction Branch Migration

The Holliday junction (HJ) suggested in 1964 by Robin Holliday is a central intermediate in homologous and site-specific recombination. Additionally, this type of DNA structure is also involved in double-stranded break repair and is a critical intermediate in restoring a stalled DNA replication process. The arrangement of two DNA molecules as an HJ assembly allows for the exchange of DNA duplexes by the process termed *branch migration*. According to schemes in Figure 2.23a, different arrangements of the junction arms are possible. Exchanging arms can be arranged in parallel (I) or antiparallel (III) conformation of the junction. Antiparallel arrangement does not allow branch migration (model III), which is the major property of HJs, although HJs with fully asymmetric sequences assemble in antiparallel conformations, as confirmed by various techniques including X-ray crystallography. AFM studies of cruciform structures that are models for naturally existing homologous HJs showed that model I is preferable, suggesting that the sequence symmetry is the factor that contributes to the structural

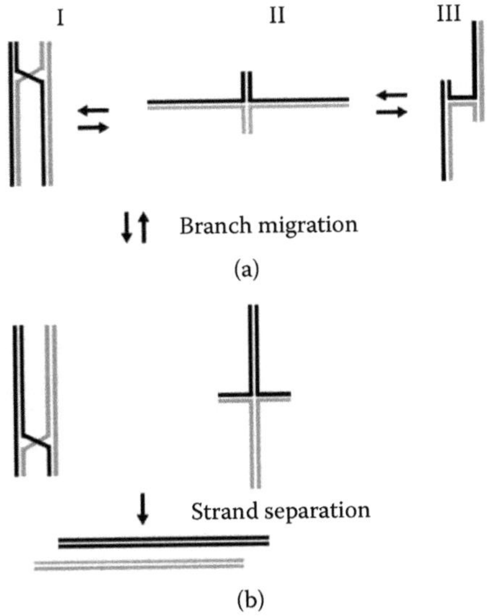

Figure 2.23 The models for Holliday junction (HJ) and branch migration. (a) Models I, II, and III correspond to the HJ conformations with the exchanging arms in the parallel arrangement, extended conformation, and antiparallel arrangement, respectively. (b) Assembly of HJ after the multiple branch migration steps. HJ in the antiparallel conformation III cannot undergo branch migration, but the other two conformations can. (c) Competition of branch migration leading to the separation of the DNA duplexes. (Reproduced from *J. Biol. Chem.*, 278, Lushnikov, A.Y., et al., DNA recombination: Holliday junctions dynamics and branch migration, 43130–43134, 2003, Copyright © (2003), with permission from Elsevier.)

dynamics of HJs (Shlyakhtenko et al. 1998, 2002a; Lyubchenko et al. 2004). Conformations I and II are capable of branch migration, and a textbook model suggested 40 years ago (Sigal and Alberts 1972) utilizes the parallel orientation of exchanging strands (model I). The attractive feature of this model is that HJ migration is accompanied by rotatory diffusion of the DNA duplexes along the major helix (Meselson 1972). This view has been challenged with an alternative model suggesting that an extended configuration of the junction (model II) is more appropriate for spontaneous movement of the HJ (Panyutin and Hsieh 1994; Panyutin et al. 1995; Biswas et al. 1998). Note that protein-mediated branch migration with the use of *E. coli* RuvAB complexes utilizes this pathway in which RuvA stabilizes extended conformation of the junction (Shlyakhtenko et al. 2000a) and two RuvB helicase assemblies bind to opposite arms of the junction to accomplish branch migration. Could it be that protein-mediated branch migration of HJ and spontaneous branch migration follow different mechanisms? Direct testing of the model of spontaneous branch migration of HJ was performed by Lushnikov et al. (2003), in which homologous HJ capable of branch migration as shown in Figure 2.23 was assembled from two DNA molecules, and its dynamics was visualized by the use of time-lapse AFM.

The junction was assembled in the presence of Mg^{2+} cations that almost terminate the branch migration process. In time-lapse AFM experiments, the HJ sample was injected into an AFM flow cell mounted on APS-mica and imaged with an AFM continuously without drying of the sample. A selected area with unambiguously identified HJs was scanned continuously, and branch migration was induced by removal of Mg^{2+} cations by the buffer change without interrupting the scanning. Few frames taken from a large set of continuous frames are shown in Figure 2.24. The initially folded arms (frame a) subsequently move to an extended conformation (frame b). One of the short arms is barely seen in frame c due to branch migration resulting in the dissociation of the molecule into two linear strands, as imaged in frame d. No branch migration was observed if the junction remained folded, and statistical analysis of the AFM data was done to conclude that unfolding of the junction with the formation of extended conformation (model II) is needed for HJ migration. The relatively rapid dissociation of the junction in such a way that the branch migration is observed in two adjacent time-lapse frames (c and d in Figure 2.24) suggests that the branch migration process is considerably faster than the AFM scanning process in which two adjacent frames were detected in the 1 minute interval. It was confirmed by Karymov et al. (2005) that branch migration with the base-pair resolution was studied with the single-molecule Förster resonance energy transfer (FRET) experiments. The authors showed that HJ branch migration is a step-wise process in which spontaneous migration is interrupted by the junction folding for dozens of milliseconds. The branch migration step is estimated as fast as 100 μs, and one branch migration hop can be larger than one base pair, a feature accelerating the entire branch migration process. It is remarkable that AFM has the capability to reveal the structural dynamics of HJ, which is not amenable to any other current structural technique.

2.7.2 Dynamics of Protein–DNA Complexes

The time-lapse AFM technique was successfully applied by a number of groups actively working with AFM of biological samples. The joint effort of the Hansma and Bustamante groups demonstrated the ability of time-lapse AFM to image the assembly process of RNA polymerase-DNA complexes (Guthold et al. 1994). In the later publication, these groups visualized the sliding of the polymerase along DNA

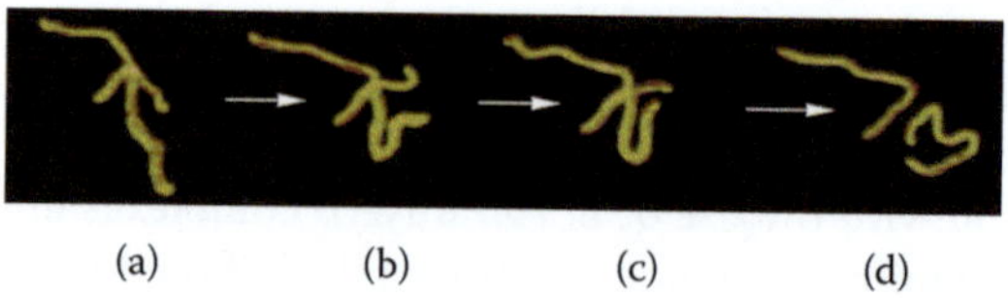

Figure 2.24 Time-lapse AFM images (a–d) illustrating branch migration of the Holliday junction. (Reproduced from *J. Biol. Chem.*, 278, Lushnikov, A.Y., et al., DNA recombination: Holliday junctions dynamics and branch migration, 43130–43134, 2003, Copyright © (2003), with permission from Elsevier.)

(Kasas et al. 1997) and showed that diffusion of RNA polymerase along DNA constituted a mechanism for accelerated promoter location. The time-lapse experiments in Guthold et al. (1999a) showed that, in addition to sliding, an enzyme can jump from one site to another. Credit in these studies should be given to substantial improvements in the DNA immobilization procedure. They modified the cation-assisted immobilization technique that provided a loose DNA binding to permit the molecule to move relatively freely on the surface. These conflicting requirements were reconciled by the use of different kinds of divalent cations. The real-time visualization of enzymatic degradation with a nuclease was performed by Bezanilla and colleagues (1994). The acceleration of the data acquisition time was made by use of the phase-imaging mode, which made it possible to visualize DNA replication by Sequenase in real time (Argaman et al. 1997). It was assumed that DNA segments between the anchoring points on the surface move freely over the surface and are available for protein interaction.

2.7.3 Structure and Dynamics of Chromatin Studied with AFM

DNA in most eukaryotic cells is tightly packed into chromatin fibril that further condensed into a chromosome. Schematically, the process of the DNA assembly is shown in Figure 2.25a. The nucleosome core particle (NCP) is a fundamental unit of chromatin structure and dynamics, and defines chromatin function. The NCP is a compact particle with 147 bp of DNA wrapped around an octameric core, composed of histones H2A, H2B, H3, and H4. The NCP structure is stabilized by electrostatic interactions, specific hydrogen bonds, and salt bridges, identified by the analysis of the high-resolution crystal structure of NCPs (Luger et al. 1997). The high stability of nucleosome particles leads to the question of how the DNA within the nucleosome can be accessed by regulatory proteins for transcription, DNA replication, recombination, and repair.

Mononucleosomes are typical experimental systems used to study nucleosome dynamics. A series of pioneering papers by Widom and collaborators (Li and Widom 2004; Li et al. 2005; Tims and Widom 2007) reported on the use of fluorescence and enzymatic approaches to probe nucleosome dynamics. These studies showed that the local opening of nucleosomes occurred in the millisecond time scale, but questions concerning the size of the openings, their locations relative to the nucleosomal ends, and the model by which the DNA segments become unfolded, required other approaches. Time-lapse AFM imaging was applied to directly visualize dynamics of mononucleosomes (Shlyakhtenko et al. 2009a). The DNA template designed for these studies was a 353 bp DNA fragment containing the 147 bp nucleosome positioning Widom-601 sequence (Li and Widom 2004), flanked with two regions of different lengths, 79bp and 127bp. The position of the nucleosome in this design is determined by the length measurements of the DNA flanks. Figure 2.25b shows AFM images of mononucleosomes assembled on the 353 bp DNA template on which the nucleosome is seen as a bright particle located between the DNA flanks. Upon DNA wrapping, the lengths of the arms gradually decrease, while the size of the nucleosome core increases. DNA wrapping is accompanied by a change in the interarm angle. Both parameters were used to determine the number of DNA turns in the

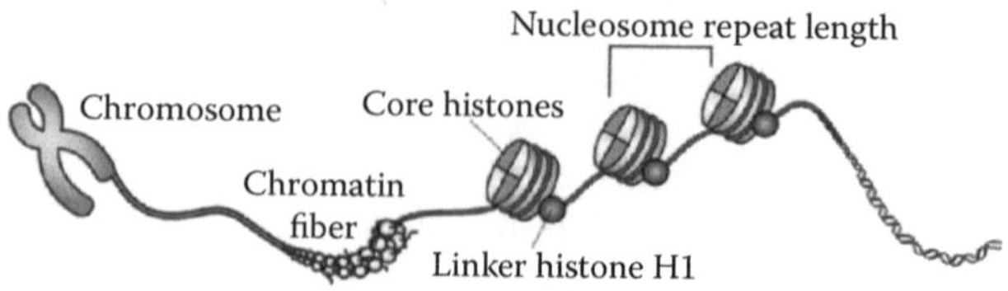

(a)

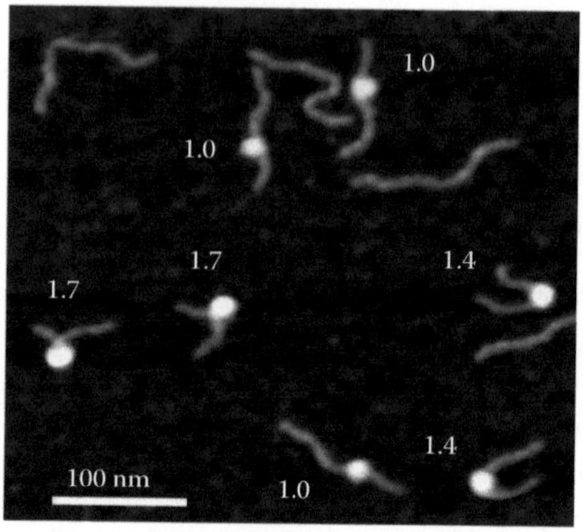

(b)

Figure 2.25 (a) Different levels of chromatin starting from the DNA duplex (on the right), the nucleosome assembly around the histone core (middle), the nucleosome fibril assembly into a higher-order chromatin fiber and the chromosome assembly (left). (Reprinted by permission from Macmillan Publishers Ltd., *Nature Med.*, Jakovcevski, M., and Akbarian, S., 18, 1194–1204, 2012, Copyright (2012).) (b) AFM image taken in air of reconstituted mononucleosomes. Particles marked 1, 2, and 3 are nucleosomes with ~1.7, ~1.4, and ~1 turns of DNA wrapped around the histone core particle. (Reproduced with permission from Shlyakhtenko, L.S., et al., *Biochemistry*, 48, 7842–7848, 2009. Copyright (2009) American Chemical Society.)

NCP particle. The numbers in the figure indicate the number of DNA turns for each particle. This value varied between 1 and 1.7 turns indicating the heterogeneity of nucleosomes by the number of DNA turns suggesting that the mononucleosome particles are dynamic. This hypothesis was tested by time-lapse imaging over nucleosomes.

The imaging frames illustrating the dynamics of one selected NCP particle with ~2 DNA turns are shown in Figure 2.26a containing selected frames of the dataset obtained by Shlyakhtenko and colleagues (2009a). Initially, a fully wrapped NCP (frame 1) slightly unwraps (frames 1 and 2); the unwrapping process is more evident in frame 3. The NCP retains its geometry over three frames in a row (frames 3, 4, and 5). Between frames 5–6, the NCP loosens and

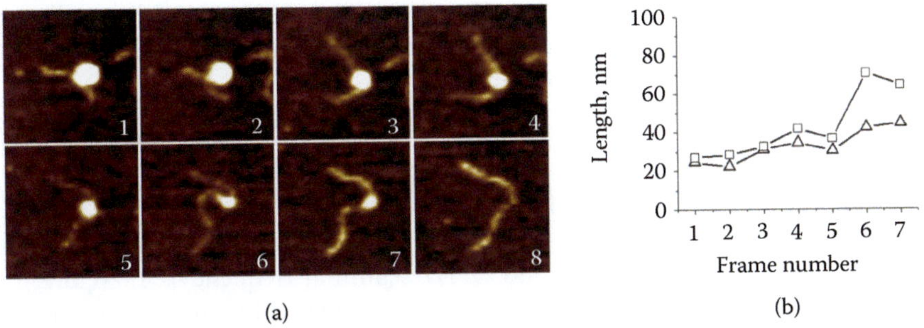

Figure 2.26 Time-lapse AFM visualization of the nucleosome unwrapping process. (a) Consecutive AFM images of nucleosomes taken during continuous scanning in the buffer. Each frame size is 200 nm. (b) Dependence of arm lengths on the frame number. (Reproduced with permission from Shlyakhtenko, L.S., et al., *Biochemistry*, 48, 7842–7848, 2009. Copyright (2009) American Chemical Society.)

unwraps again in frame 6. It remains unchanged in frame 7 and finally undergoes full dissociation in frame 8. The length measurement (Figure 2.26b) shows that the nucleosome dissociation is accompanied by the unwrapping of both DNA flanks, although during the final stages (frames 6 and 7), the process is asymmetric with substantial elongation of one arm compared to another. Similar observations led to the conclusion that the nucleosome undergoes spontaneous dissociation, although the specific path can vary from particle to particle. The example described illustrates a nonuniform dissociation pathway; however, examples of uniform time-dependent dissociation have been reported as well (Shlyakhtenko et al. 2009a).

The AFM time-lapse experiments directly illustrate a highly dynamic feature of mononucleosomes, suggesting that nucleosomes are rather unstable. This finding is in line with direct measurements of the NCP stability in diluted solutions in experiments in which the ratio of intact nucleosomes as a function of time was measured by AFM imaging (Menshikova et al. 2011). At the same time, we need to take into account the effect of the surface used for the AFM experiment. Although APS-mica has a low charge density and has little effect on the ion distribution near the surface (Lyubchenko and Shlyakhtenko 1997), the interaction of DNA with the surface favors the unwrapping process. It was hypothesized by Miyagi et al. (2011) that electrostatics can play a role in chromatin unfolding by remodeling factors. Specifically it was posited that the surface of chromatin remodeling factors accommodating the nucleosome would be positively charged to facilitate nucleosome unfolding, and a study from the Kornberg lab (Lorch et al. 2010) supports this hypothesis. The effect of local surface properties on nucleosome dynamics can be generalized to the chromatin level. The interaction of chromatin with nuclear membranes and other components of the nuclear matrix are important factors that modulate chromatin structure and function.

Designing the surfaces with various characteristics would mimic the membrane structure, and the AFM as a topographic technique is an ideal method to characterize these interactions.

2.8 High-Speed AFM

As previously described, AFM images are reconstructed through point-by-point probing of the surface topography. This operation is typically performed in the range of one scan line per second (2 Hz scanning frequency). Therefore, the acquisition of the entire image comprising 512 lines (512 data points per line) takes ~4 minutes. For dry samples this slowness of the AFM operation is a matter of inconvenience and in fact is not a serious bottleneck of the entire data acquisition process. However, a low data acquisition rate of AFM is a serious deficiency if AFM is applied to imaging of the majority of biological processes. For example, the translocation of RNA polymerase along DNA occurs on the second time scale; therefore, the observation on the minute time scale reveals only a few percent of the events.

Developments of AFM that enabled high data acquisition speed were made by several groups. The first fast scanning AFM was made in the group of the AFM co-inventor Quate (Manalis et al. 1996), in which an integrated active actuator was used to achieve high scanning speed. However, a high stiffness of active probes did not allow for the implementation of this idea into the imaging of soft biological systems. The breakthrough in the development of the high-speed (HS) AFM for biological application was made after implementation of the short cantilever idea proposed by Hansma (Walters et al. 1996). The need for tip modification and the importance of the small cantilever idea can be illustrated using the equation for the resonant frequency of the cantilever and the mass:

$$f_0 \frac{1}{2\pi}\sqrt{\frac{k}{m}} \qquad k = \frac{Ewt^3}{4l^3} \tag{2.3}$$

where k is the spring constant, E is Young's modulus, t is thickness, l is length, w is width, and m is the effective mass of the cantilever.

To achieve a high scanning speed, we need to increase the cantilever frequency or/and decrease the cantilever mass. The increase of the first parameter is undesirable, as imaging of biological samples requires soft cantilevers. Thus, decrease of the mass is the only way to achieve the goal. However, to increase the resonant frequency 100 times, the cantilever mass should be 10,000 times less compared to an original design, which is a technically challenging goal. The initial HS-AFM design made in the Hansma group provided data acquisition with the rate of 0.5 frames/s, which was 100 times faster than the rate of a conventional AFM (Fantner et al. 2006; Hansma et al. 2006).

The short cantilever idea was implemented by Ando in his first design of an HS-AFM instrument capable of operating with the subsecond image acquisition speed (Ando et al. 2001, 2008, 2014, 2014a, Box 2.4, Figure 2.27). Key issues for

Box 2.4 High-Speed AFM

Given the small size of the cantilever (3; i.e., 6–10 μm long, 2 μm wide, and 90 nm thick), a special optics was used to measure deflection of the cantilever, enabling detection of vertical movement of the cantilevers with a subnanometer accuracy. The objective lenses (9) focus the laser spot on the rear side of the cantilever (3); the reflected laser beam is collected and collimated with the same objective lenses (9). The incident and reflected light beams are separated using a polarized beam splitter (12) and a quarter-wavelength plate (11). Then, these light beams are reflected onto the split photodiode (15). The cantilever (3) is attached to a transparent glass plate via a short cantilever holder glued to the glass plate. The rear side of the cantilever faces downward, and the tip faces upward to facilitate incorporation of the optics. The scanner (1) with a sample stage (2) is mounted facing downward so the specimen is supported at the bottom of the sample stage and inverted to the level of the cantilever probe. The angle between the tip and the sample is defined by the angular position of the scanner to the sample stage. The resonant frequency and the spring constants of the HS-AFM cantilevers are in the range of 600 kHz–1.2 MHz in water and 0.1–0.2 N/m, respectively. Such cantilevers are commercially available (e.g., AC-10; Olympus, Tokyo, Japan).

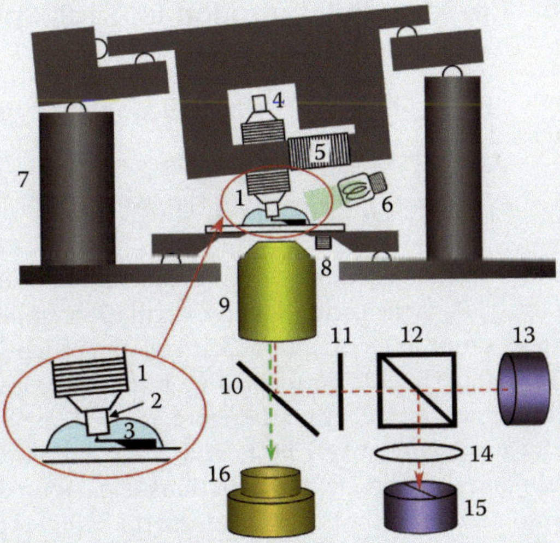

Figure 2.27 HS-AFM head integrated with an inverted type of optical microscope. Z-piezo scanner (1), sample stage (2), cantilever (3), counterweight (4), x-piezo scanner (5), illuminator (6), stepper motor (7), piezo for exciting cantilever (8), objective lens (9), dichroic mirror (10), quarter-wave plate (11), polarization beam splitter (12), laser diode (13), focusing lens (14), split photodiode (15), and CCD camera (16). (Reproduced from Lyubchenko, Y.L. et al, *Methods* 54:274–283, 2011. With permission.)

high-speed AFM realized in the Ando design are as follows: (1) small cantilevers with a resonant frequency above several hundred kHz and a small spring constant, (2) a high-speed scanner with a resonant frequency that matched the cantilever characteristics, (3) active damping techniques to suppress mechanical vibrations of the scanner, and (4) fast feedback control. The current HS-AFM instrument can capture images with the video rate without significant disturbance of weak biomolecular interactions (Ando 2014b) (see Box 2.4 for the instrument layout).

The drive amplitude issue is especially important for the HS-AFM due to the elevated scanning speed. Therefore, operation of the instrument at low drive amplitude received careful attention in the instrument design that operates with amplitudes an order of magnitude lower than traditional AFM. The analysis of the effect of the tip on the sample for HS-AFM was done by Miyagi and colleagues (2011). The HS-AFM operates in tapping mode; therefore, we need to consider the effect of two major factors: the hitting of the sample with an oscillating tip and displacement of the sample during scanning. The energy transferred by the tip to the sample is proportional to the square of the oscillation amplitude. HS-AFM operates with an amplitude of ~1 nm, so the estimates (see Box 2.5) demonstrate that the overall energy induced by the tip is equivalent to the increase of temperature

Box 2.5 Tip-Sample Interaction for High-Speed AFM (see Miyagi et al. 2011)

The maximum tapping force generated by an oscillating tip to the sample (f_{max}) is expressed as

$$f_{max} = k_c(A_0^2 - A_s^2)^{1/2}/Q_c, \tag{2.4}$$

where k_c and Q_c are the cantilever's spring constant and quality factor in water, respectively; A_0 is the cantilever free oscillation amplitude; and A_s is the amplitude set point to be maintained as constant during scanning. For typical HS-AFM experiments performed with the EBD tips, $k_c = 0.2$ N/m, $Q_c = 2$, $A_0 = 1$ nm, and $A_s = 0.9$ nm; therefore, $f_{max} = 22$ pN. Although such a magnitude of force is relatively large and is comparable with dynamic protein-protein and protein–DNA interactions (e.g., Tong et al. 2013), one should take into account a short period of the tip-sample contact. Indeed, small cantilevers used for HS-AFM have a resonant frequency of 1.2 MHz in water; therefore, the force acting time is approximately 100 ns suggesting that impulse is really small. It is also instructive to use this estimate to evaluate possible displacement of the sample during the tapping touch. In a typical HS-AFM experiment, the scan velocity in the *x*-direction is at most 1.6 mm/s (under the condition of 33 frames/s, 250 nm scan range, and 100 scan lines). While the tip is in contact with a sample (~100 ns), the sample stage moves in the *x*-direction only by 0.16 nm. This value is comparable to the size of an atom, so the lateral displacement of the sample by the scanning tip is negligible.

The following expression can be used to estimate the amount of energy transferred to a sample by oscillating tip:

$$\Delta E = 0.5 \times k_c A_0^2 - 0.5 \times k_c A_s^2 \quad (2.5)$$

For $A_s = 0.9 \times A_0$, we obtain $\Delta E = 0.5 \times k_c A_0^2 \times 0.19 = 0.095 \times k_c A_0^2$, and for $k_c = 0.2$ N/m and $A_0 = 1$ nm, ΔE is 1.9×10^{-20} Joule.

If the thermal energy distributed to one freedom is $0.5 \times k_B T$, where k_B is Boltzmann constant and T is the absolute temperature, at $T = 300$ K, the thermal energy is ~2×10^{-21} Joule. Therefore, the energy provided by the tip is ~$9.5 \times (0.5 \times k_B T)$. The temperature increase ΔT by each atom is 9.5 $T/3N$, where each atom is assumed to have three freedoms. If $N = 100$, the temperature increase is 0.032 $T = 0.032 \times 300 = 9.6$ K. Given that the atom's dynamics is in the range of picoseconds, acquired energy during the 100 ns contact time is quickly distributed to more freedoms. We also need to take into account that HS-AFM operates in an aqueous environment, so the sample is in the water bath that keeps the temperature constant.

by ~9.6 K only. Given the fact that this energy is distributed over the entire system, including water molecules in the vicinity of the sample, this effect is low. Another effect of the tip is the displacement of the sample. Because of the high oscillation frequency, the time during which the tip contacts the sample is ~100 ns; therefore, during this period of time the sample stage moves in the x-direction only by 0.16 nm. This value is comparable to the size of an atom, so the lateral displacement of the sample by the scanning tip is negligible.

The dragging effect was investigated by Kobayashi and colleagues (2007), who measured the movement rates of DNA segment along the tip scanning direction and perpendicular to it of the HS-AFM instrument. A difference has not been found, suggesting that the lateral force applied to the sample does not increase essentially with the increasing scan rate. Additional studies were conducted by Suzuki et al. (2010) who calculated the elastic bending energies and fluctuations of single molecule conformations to assess the effect of the probe.

Recent review articles describe the current advances of the application of HS-AFM to biological problems (Ando 2014, 2014b). In the following, we briefly describe the results of two papers in which interesting biological problems were targeted using this unique nanoimaging approach.

2.8.1 Nanoscale Visualization of Myosin V Translocation Along the Actin Fibril

Myosin V belongs to a family of ATP-dependent motor proteins playing key roles in muscle contraction as well as other motility processes. Myosin V is critical in vesicle movement from the center of the cell to the periphery and also plays the role of a dynamic tether, retaining vesicles and organelles in the actin-rich periphery of cells. To accomplish its function, myosin V walks along actin filaments with a ~36 nm step toward the positive end of the filaments. This information was primarily obtained by the single molecule optical imaging and optical trap probing techniques that demonstrated the processive hand-over-hand movement of the motor coupled with hydrolysis of adenosine triphosphate (e.g., Sakamoto et al. 2008). Previous imaging studies of the protein studies using electron microscopy, X-ray crystallography, and nuclear magnetic resonance recording have only produced static images of the protein. By using HS-AFM, Ando and coworkers were able to simultaneously record the structure and dynamics of myosin V molecules moving along actin filaments (Kodera et al. 2010; Ando 2014b).

Results of the HS-AFM studies are summarized in Figure 2.28. A set of a few time-lapse AFM images of a selected myosin V molecule attached to actin fibrils is

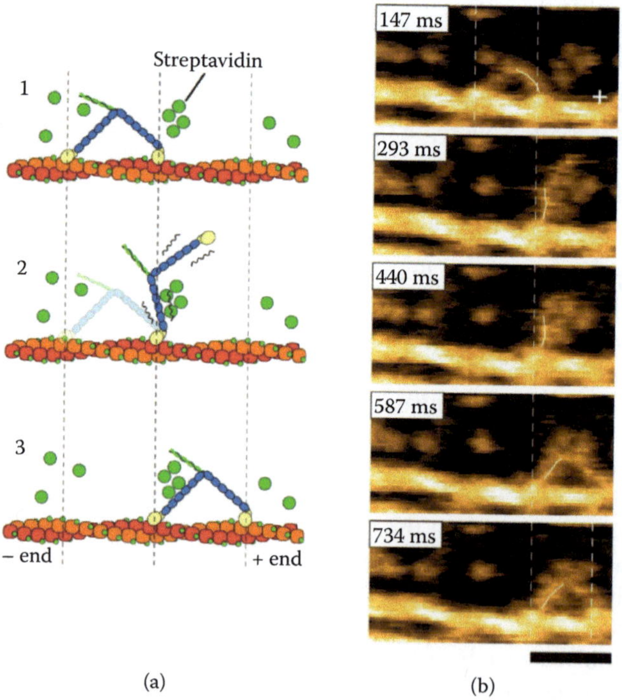

Figure 2.28 Stepping behavior of M5- HMM visualized by HS-AFM. (a) Explains HSAFM images shown in (b). Successive HS-AFM images that resolved the stepping behavior of M5-HMM in 1 μM ATP. Frame rate, 7 fps; scan area, 150 nm × 75 nm with 80 × 40 pixels; scale bar 50 nm. The swinging lever-arm is highlighted with the thin white lines. The vertical dashed lines in (a) and (b) represent the centers of mass of the motor domains, and the plus sign indicates the plus end of actin. (Reproduced from Ando, T., *Curr. Opin. Struct. Biol.*, 28, 63–68, 2014. With permission.)

shown in frame b, whereas frame a is a schematic to interpret the images. Myosin V attached to actin appears as a V-shape molecule clearly seen in the top frame of the dataset, and such a geometry is in line with EM images obtained earlier. The second frame illustrates the beginning of the protein translocation in which one arm detaches from the filament, whereas another arm remains bound to the same position on the filament. The entire image of the protein is blurry due to the dynamic feature of the system, and it remains of the same morphology at the next AFM image (440 ms). Schematic 2 in set (a) summarizes this arrangement. The next step (image four from the top; 587 ms) shows that the initially dissociated arm binds to the actin filament, restoring the V-shape of myosin V, and schematic 3 in (a) illustrates this assembly. Vertical dashed lines on both sets indicate positions of the myosin head bound to the filament. The experiments were performed in the presence of an excess of unbound streptavidin molecules appearing as small spherical particles. Hitting these streptavidin molecules decelerates the myosin walking rate allowing the authors to visualize the walking process within the temporal resolution of the instrument. The results are in a perfect coincidence with the model of myosin walking and therefore represent the first direct visualization of the important biological process at molecular-level resolution. These studies also revealed a number of dynamic properties of myosin V, such as the junction flexibility, and also highlighted a set of new problems that can be solved in future HS-AFM studies (Kodera and Ando 2014).

2.8.2 Site Search Process of Site-Specific DNA-Binding Proteins

The interaction between distant DNA regions is the key step in such fundamental genetic processes as site-specific recombination, integration, excision, and inversion of specific DNA regions within genomes (Grindley et al. 2006), as well as V(D)J recombination (Swanson 2004). This interaction is controlled by a specialized protein or protein complex that is responsible for forming a site-specific protein–DNA synaptic complex (synaptosome). In fact, the formation of a synaptic complex is a general phenomenon not limited to site-specific recombination systems. For example, a large family of DNA restriction enzymes forms synaptic complexes to facilitate further site-specific DNA cleavage (Bilcock and Halford 1999; Embleton et al. 2001; Bath et al. 2002; Gormley et al. 2002). How does the protein find both specific sites? Apparently, it finds one site and then searches for another one. What is the process? If the two sites are located in one DNA molecule, DNA loops are formed. Does looping create structural constraints? These questions were addressed in a study by Gilmore et al. (2009) in which the interaction of the EcoRII restriction enzyme belonging to the synaptosome family of site-specific DNA proteins was visualized with HS-AFM.

The time-lapse experiments showed that EcoRII slides and dissociates or associates employing 3D and 2D diffusion while searching for the first site. Unexpected behavior was found when EcoRII captured two DNA sites with one of them being specific, leading to the formation of the loop. Frame 2.5 s of Figure 2.29a shows this initial looped structure with the protein appearing as a bright blot at the strands' intersection. The images following this one illustrate the dynamic of the complex with the DNA loop. It is seen from the comparison of

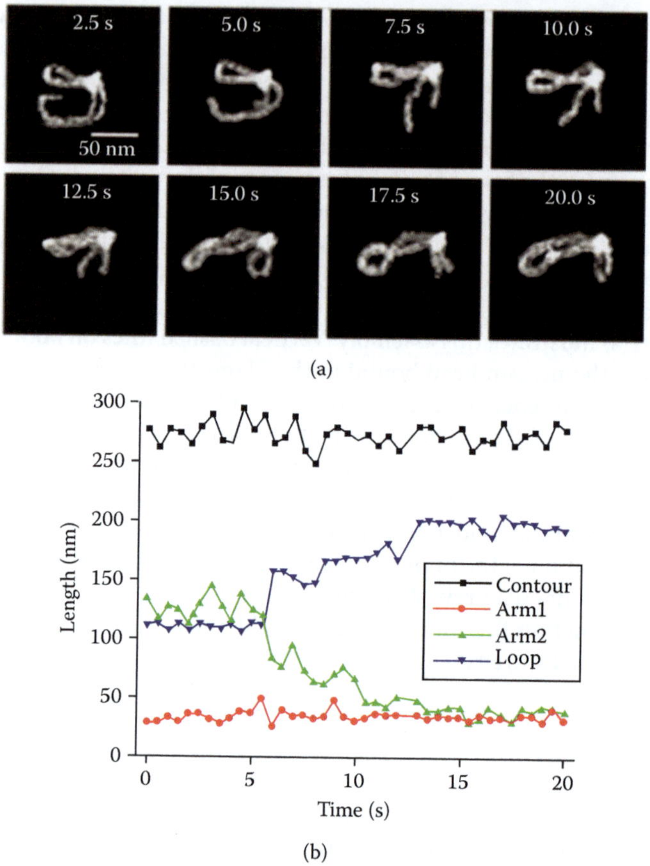

Figure 2.29 HS-AFM study of the dynamics of EcoRII-DNA complexes. (a) Individual frames shown every 2.5 s illustrate EcoRII translocation. As the long arm gradually decreases, the loop length gradually increases. (b) The dependence of DNA length on time measured in 0.5 s intervals. As the long arm gradually gets shorter, the loop length gradually increases. The contour lengths of the entire molecule and the short arm have consistent values over the entire timescale. The translocation over the length of 300 base pairs occurs within 10 s. (Reproduced from Gilmore, J.L., et al., *Biochemistry,* 48, 10492–10498, 2009. With permission.)

frames 5 s and 7.5 s that the loop increases over time and the long arm decreases in size. This process continues, so the last image (frame 20 s) shows considerable increase of the loop size. This process was analyzed by the contour length measurements (Figure 2.29b). The short arm 1 remains constant over the observation period, and the 34 nm length corresponds to the position for the EcoRII specific binding site at 100 bp from the DNA end, suggesting that the protein remains specifically bound at this recognition site the entire time. The length of arm 2 fluctuates with shortening starting after the 5 s interval, and the loop size is between 283 bp (96 nm) and 312 bp (106 nm). At about 6 s, the loop length begins to increase as arm 2 length decreases, showing translocation occurs along this arm 2. After 10 s, translocation stops, and this position corresponds to the location of another specific DNA site for EcoRII as validated by the length measurements, 34 nm or 100 bp from the end. During this process the loop becomes as large as 610 bp. Thus, the translocation process covers a distance of about 300 bp (102 nm) with the mean rate of about 30 bp/s (10.2 nm/s).

A number of conclusions emerge from these studies. First, the search process is such that EcoRII remaining bound at one DNA recognition site searches for another site by threading of the DNA filament through another DNA binding pocket of the enzyme until the enzyme finds the second recognition site and subsequently forms a stable synaptic complex. Second, although there are fluctuations of the loop size, it gradually increases as the formation of the larger loop is an entropically favorable process. Third, the protein translocation is not accompanied by its rotation around the DNA helix, as DNA supercoiling within the loop would accompany it. This finding suggests that during translocation, the protein slides along the DNA filament, maybe via small hops. These are rather unexpected findings, and future studies of other site-specific systems would answer the question of would the site search features of EcoRII and HS-AFM be an appropriate tool for that.

2.9 Conclusions: Future of AFM Topographic Studies

AFM, with its capability to image biological samples in a fully hydrated state, is coming of age. The use of time-lapse AFM to image DNA in liquid in complex with various proteins, including ligands, enables direct imaging of critical genetic processes, such as transcription, replication, and recombination. Recent advances in high-speed AFM have enabled this technique to become available to a broader biomedical community. At its current stage of development, HS AFM is capable of imaging molecular dynamics with the rate approaching video optical microscopy. However, the unique capability to provide the dynamics at the nanoscale range opens novel prospects for nanoimaging. Moreover, the current HS AFM design operates with the oscillation amplitude in the range of a nanometer, thereby dramatically minimizing the potential deformation effect of the tip. This technological advance is an important step for further improvement of the instrument, enabling AFM to operate in the least invasive mode (attractive tip-sample interaction regime). The latter will almost eliminate the tip-induced modification of the sample. The tip geometry is the major resolution-limiting factor for AFM. Recent advances in the tip manufacturing process provided the AFM practitioners with probes as sharp as a few nanometers. At the same time, there is room for improvement in instrumentation, such as high speed for large samples and sample preparation techniques. Finally, the technological advances of AFM analysis of various biological complexes can be extended to other biomolecular systems and thus will provide structural biologists pursuing single molecule approaches with novel, powerful nanoimaging tools.

References

Amirikyan BR, Vologodskii AV, Lyubchenko YL (1981) Determination of DNA cooperativity factor. *Nucleic Acids Res* 9:5469–5482.

Ando T (2014b) High-speed AFM imaging. *Curr Opin Struct Biol* 28:63–68.

Ando T, Kodera N, Takai E, Maruyama D, Saito K, Toda A (2001) A high-speed atomic force microscope for studying biological macromolecules. *Proc Natl Acad Sci U S A* 98:12468–12472.

Ando T, Uchihashi T, Fukuma T (2008) High-speed atomic force microscopy for nano-visualization of dynamic biomolecular processes. *Progr Surf Sci* 83:337–437.

Ando T, Uchihashi T, Scheuring S (2014) Filming biomolecular processes by high-speed atomic force microscopy. *Chem Rev* 114:3120–3188.

Argaman M, Golan R, Thomson NH, Hansma HG (1997) Phase imaging of moving DNA molecules and DNA molecules replicated in the atomic force microscope. *Nucleic Acids Res* 25:4379–4384.

Bath AJ, Milsom SE, Gormley NA, Halford SE (2002) Many type IIs restriction endonucleases interact with two recognition sites before cleaving DNA. *J Biol Chem* 277:4024–4033.

Berquand A, Xia N, Castner DG, Clare BH, Abbott NL, Dupres V, Adriaensen Y, Dufrene YF (2005) Antigen binding forces of single antilysozyme Fv fragments explored by atomic force microscopy. *Langmuir* 21:5517–5523.

Bezanilla M, Drake B, Nudler E, Kashlev M, Hansma PK, Hansma HG (1994) Motion and enzymatic degradation of DNA in the atomic force microscope. *Biophys J* 67:2454–2459.

Bezanilla M, Manne S, Laney DE, Lyubchenko YL, Hansma HG (1995) Adsorption of DNA to Mica, Silylated Mica, and minerals: Characterization by atomic force microscopy. *Langmuir* 11:655–659.

Bianco PR (2015) I came to a fork in the DNA and there was RecG. *Prog Biophys Mol Biol* 117:166–173.

Bikard D, Loot C, Baharoglu Z, Mazel D (2010) Folded DNA in action: Hairpin formation and biological functions in prokaryotes. *Microbiol Mol Biol Rev* 74:570–588.

Bilcock DT, Halford SE (1999) DNA restriction dependent on two recognition sites: Activities of the SfiI restriction-modification system in Escherichia coli. *Mol Microbiol* 31:1243–1254.

Binnig G, Quate CF, Gerber C (1966) Atomic force microscope. *Phys Rev Lett* 56:930–933

Binnig G, Rohrer H, Gerber C, Weibel E. (1982) Tunneling through a controllable vacuum gap. *Appl Phys Lett* 40:178–180.

Biswas I, Yamamoto A, Hsieh P (1998) Branch migration through DNA sequence heterology. *J Mol Biol* 279:795–806.

Bustamante C, Erie D, Keller D (1994) Biochemical and structural applications of scanning force microscopy. *Curr Opinion in Struct Biol* 4:750–760.

Bustamante C, Rivetti C (1996) Visualizing protein-nucleic acid interactions on a large scale with the scanning force microscope. *Annu Rev Biophys Biomol Struct* 25:395–429.

Bustamante C, Rivetti C, Keller DJ (1997) Scanning force microscopy under aqueous solutions. *Curr Opin Struct Biol* 7:709–716.

Bustamante C, Vesenka J, Tang CL, Rees W, Guthold M, Keller R (1992) Circular DNA molecules imaged in air by scanning force microscopy. *Biochemistry* 31:22–26.

Butt H-J, Dowing KH, Hansma PK (1990) Imaging the membrane protein bacteriorhodopsin with the atomic force microscope. *Biophys J* 58:1473–1480.

Drake B, Prater CB, Weisenhorn AL, Gould SA, Albrecht TR, Quate CF, Cannell DS, Hansma HG, Hansma PK (1989) Imaging crystals, polymers, and processes in water with the atomic force microscope. *Science* 243:1586–1589.

Dulbecco R, Vogt M (1963) Evidence for a ring structure of polyoma virus DNA. *Proc Natl Acad Sci U S A* 50:236–243.

Dzyaloshinskii I, Lifshitz E, Pitaevskii LP (1961) General theory of van der Waals forces. *Sov Phys Uspekhi* 4:153–176.

Embleton ML, Siksnys V, Halford SE (2001) DNA cleavage reactions by type II restriction enzymes that require two copies of their recognition sites. *J Mol Biol* 311:503–514.

Fantner GE, Schitter G, Kindt JH, Ivanov T, Ivanova K, Patel R, Holten-Andersen N, Adams J, Thurner PJ, Rangelow IW, Hansma PK (2006) Components for high speed atomic force microscopy. *Ultramicroscopy* 106:881–887.

Freund J, Halbritter J, Horber JK (1999) How dry are dried samples? Water adsorption measured by STM. *Microsc Res Tech* 44:327–338.

Gilmore JL, Suzuki Y, Tamulaitis G, Siksnys V, Takeyasu K, Lyubchenko YL (2009) Single-molecule dynamics of the DNA-EcoRII protein complexes revealed with high-speed atomic force microscopy. *Biochemistry* 48:10492–10498.

Gormley NA, Hillberg AL, Halford SE (2002) The type IIs restriction endonuclease BspMI is a tetramer that acts concertedly at two copies of an asymmetric DNA sequence. *J Biol Chem* 277:4034–4041.

Grindley ND, Whiteson KL, Rice PA (2006) Mechanisms of Site-Specific Recombination. *Annu Rev Biochem* 75:567–605.

Gross L, Mohn F, Moll N, Schuler B, Criado A, Guitian E, Pena D, Gourdon A, Meyer G (2012) Bond-order discrimination by atomic force microscopy. *Science* 337:1326–1329.

Guthold M, Bezanilla M, Erie DA, Jenkins B, Hansma HG, Bustamante C (1994) Following the assembly of RNA polymerase-DNA complexes in aqueous solutions with the scanning force microscope. *Proc Natl Acad Sci U S A* 91:12927–12931.

Guthold M, Zhu X, Rivetti C, Yang G, Thomson NH, Kasas S, Hansma HG, Smith B, Hansma PK, Bustamante C (1999a) Direct observation of one-dimensional diffusion and transcription by Escherichia coli RNA polymerase. *Biophys J* 77:2284–2294.

Guthold M, Zhu X, Rivetti C, Yang G, Thomson NH, Kasas S, Hansma HG, Smith B, Hansma PK, Bustamante C (1999b) Direct observation of one-dimensional diffusion and transcription by Escherichia coli RNA polymerase. *Biophys J* 77:2284–2294.

Hamaker HC (1937) The London – Van der Waals attraction between spherical particles. *Physica* 4:1058.

Hansma HG, Hoh JH (1994) Biomolecular imaging with the atomic force microscope. *Annu Rev Biophys Biomol Struct* 23:115–139.

Hansma PK, Schitter G, Fantner GE, Prater C (2006) Applied physics. High-speed atomic force microscopy. *Science* 314:601–602.

Heinisch JJ, Dupres V, Wilk S, Jendretzki A, Dufrene YF (2010) Single-molecule atomic force microscopy reveals clustering of the yeast plasma-membrane sensor Wsc1. *PLoS One* 5:e11104.

Henderson E, Haydon PG, Sakaguchi DS (1992) Actin filament dynamics in living glial cells imaged by atomic force microscopy. *Science* 257:1944–1946.

Higgins NP, Vologodskii AV (2015) Topological behavior of plasmid DNA. *Microbiol Spectr* 3(2).

Hoh JH, Sosinsky GE, Revel JP, Hansma PK (1993) Structure of the extracellular surface of the gap junction by atomic force microscopy. *Biophys J* 65:149–163.

Josephs EA, Zheng T, Marszalek PE (2015) Atomic force microscopy captures the initiation of methyl-directed DNA mismatch repair. *DNA Repair* 35:71–84.

Karymov M, Daniel D, Sankey OF, Lyubchenko YL (2005) Holliday junction dynamics and branch migration: Single-molecule analysis. *Proc Natl Acad Sci U S A* 102:8186–8191.

Kasas S, Thomson NH, Smith BL, Hansma HG, Zhu X, Guthold M, Bustamante C, Kool ET, Kashlev M, Hansma PK (1997) Escherichia coli RNA polymerase activity observed using atomic force microscopy. *Biochemistry* 36:461–468.

Kato M, McAllister CJ, Hokabe S, Shimizu N, Lyubchenko YL (2002) Structural heterogeneity of pyrimidine/purine-biased DNA sequence analyzed by atomic force microscopy. *Eur J Biochem* 269:3632–3636.

Keller RW, Keller DJ, Bear D, Vasenka J, Bustamante C (1992) Electrodeposition procedure of E. coli RNA polymerase onto gold and deposition of E. coli RNA polymerase onto mica for observation with scanning force microscopy. *Ultramicroscopy* 42–44 (Pt B):1173–1180.

Klinov D, Dwir B, Kapon E, Borovok N, Molotsky T, Kotlyar A (2007) High-resolution atomic force microscopy of duplex and triplex DNA molecules. *Nanotechnology* 18:225102.

Kobayashi M, Sumitomo K, Torimitsu K (2007) Real-time imaging of DNA-streptavidin complex formation in solution using a high-speed atomic force microscope. *Ultramicroscopy* 107:184–190.

Kodera N, Ando T (2014) The path to visualization of walking myosin V by high-speed atomic force microscopy. *Biophys Rev* 6:237–260.

Kodera N, Yamamoto D, Ishikawa R, Ando T (2010) Video imaging of walking myosin V by high-speed atomic force microscopy. *Nature* 468:72–76.

Lennard-Jones JE (1931) Cohesion. *Proceedings of the Physical Society* 43:461–482.

Li G, Levitus M, Bustamante C, Widom J (2005) Rapid spontaneous accessibility of nucleosomal DNA. *Nat Struct Mol Biol* 12:46–53.

Li G, Widom J (2004) Nucleosomes facilitate their own invasion. *Nat Struct Mol Biol* 11:763–769.

Lorch Y, Maier-Davis B, Kornberg RD (2010) Mechanism of chromatin remodeling. *Proceedings of the National Academy of Sciences of the United States of America* 107:3458–3462.

Luger K, Mader A, Richmond R, Sargent D, Richmond T (1997) Crystal structure of the nucleosome core particle at 2.8Å resolution. *Nature* 389:251–260.

Lushnikov AY, Bogdanov A, Lyubchenko YL (2003) DNA recombination: Holliday junctions dynamics and branch migration. *The Journal of Biological Chemistry* 278:43130–43134.

Lushnikov AY, Potaman VN, Oussatcheva EA, Sinden RR, Lyubchenko YL (2006) DNA Strand Arrangement within the sfiI-DNA complex: Atomic force microscopy analysis. *Biochemistry* 45:152–158.

Lyubchenko Y, Shlyakhtenko L, Harrington R, Oden P, Lindsay S (1993a) Atomic force microscopy of long DNA: Imaging in air and under water. *Proc Natl Acad Sci U S A* 90:2137–2140.

Lyubchenko YL (2004) DNA structure and dynamics: An atomic force microscopy study. *Cell Biochem Biophys* 41:75–98.

Lyubchenko YL (2011) Preparation of DNA and nucleoprotein samples for AFM imaging. *Micron* 42:196–206.

Lyubchenko YL, Gall AA, Shlyakhtenko LS (2001) Atomic force microscopy of DNA and protein–DNA complexes using functionalized mica substrates. *Methods Mol Biol* 148:569–578.

Lyubchenko YL, Gall AA, Shlyakhtenko LS, Harrington RE, Jacobs BL, Oden PI, Lindsay SM (1992a) Atomic force microscopy imaging of double stranded DNA and RNA. *J Biomol Struct Dynam* 10:589–606.

Lyubchenko YL, Jacobs BL, Lindsay SM (1992b) Atomic force microscopy of reovirus dsRNA: A routine technique for length measurements. *Nucleic Acids Res* 20:3983–3986.

Lyubchenko YL, Jacobs BL, Lindsay SM, Stasiak A (1995) Atomic force microscopy of nucleoprotein complexes. *Scanning Microsc* 9:705–724; discussion 724–707.

Lyubchenko YL, Oden PI, Lampner D, Lindsay SM, Dunker KA (1993b) Atomic force microscopy of DNA and bacteriophage in air, water and propanol: The role of adhesion forces. *Nucleic Acids Res* 21:1117–1123.

Lyubchenko YL, Shlyakhtenko LS (1997) Visualization of supercoiled DNA with atomic force microscopy in situ. *Proc Natl Acad Sci U S A* 94:496–501.

Lyubchenko YL, Shlyakhtenko LS (2009) AFM for analysis of structure and dynamics of DNA and protein–DNA complexes. *Methods* 47:206–213.

Lyubchenko YL, Shlyakhtenko LS, Gall AA (2009) Atomic force microscopy imaging and probing of DNA, proteins, and protein DNA complexes: Silatrane surface chemistry. *Methods Mol Biol* 543:337–351.

Lyubchenko YL, Shlyakhtenko LS, Potaman VP, Sinden RR. (2002) Global and Local DNA structure and dynamics. Single molecule studies with AFM. *Microsc Microanal* 8:170–171.

Manalis SR, Minne SC, Quate CF (1996) Atomic force microscopy for high speed imaging using cantilevers with an integrated actuator and sensor. *Appl Phys Lett* 68:871–873.

Martin Y, Williams CC, Wickramasinghe HK (1987) Atomic force microscopy-force mapping and profiling on a sub 100A scale. *J Appl Phys* 61:4723–4729.

McGlynn P, Lloyd RG (2002) Genome stability and the processing of damaged replication forks by RecG. *Trends Genet* 18:413–419.

Menshikova I, Menshikov E, Filenko N, Lyubchenko YL (2011) Nucleosomes structure and dynamics: Effect of CHAPS. *Int J Biochem Mol Biol* 2:129–137.

Meselson M (1972) Formation of hybrid DNA by rotary diffusion during genetic recombination. *J Mol Biol* 71:795–798.

Mikheikin AL, Lushnikov AY, Lyubchenko YL (2006) Effect of DNA supercoiling on the geometry of holliday junctions. *Biochemistry* 45:12998–13006.

Mirkin SM (2008) Discovery of alternative DNA structures: A heroic decade (1979–1989). *Front Biosci* 13:1064–1071.

Mironov LV (2004) *Fundamentals of scanning probe microscopy.* Nizhny Novgorod: Russian Academy of Sciences.

Miyagi A, Ando T, Lyubchenko YL (2011) Dynamics of nucleosomes assessed with time-lapse high-speed atomic force microscopy. *Biochemistry* 50:7901–7908.

Modrich P (2006) Mechanisms in eukaryotic mismatch repair. *J Biol Chem* 281:30305–30309.

Mou J, Czajkowsky DM, Zhang Y, Shao Z (1995) High-resolution atomic-force microscopy of DNA: The pitch of the double helix. *FEBS Lett* 371:279–282.

Muller DJ (2008) AFM: A nanotool in membrane biology. *Biochemistry* 47:7986–7998.

Muller DJ, Engel A (2007) Atomic force microscopy and spectroscopy of native membrane proteins. *Nature protocols* 2:2191–2197.

Nagami F, Zuccheri G, Samori B, Kuroda R (2002) Time-lapse imaging of conformational changes in supercoiled DNA by scanning force microscopy. *Analytical Biochemistry* 300:170–176.

Ohnishi S, Hara M, Furuno T, Sasabe H (1992) Direct visualization of polypeptide shell of ferritin molecule by atomic force microscopy. *Biophys J* 63:1425–1431.

Panyutin IG, Biswas I, Hsieh P (1995) A pivotal role for the structure of the Holliday junction in DNA branch migration. *Embo J* 14:1819–1826.

Panyutin IG, Hsieh P (1994) The kinetics of spontaneous DNA branch migration. *Proc Natl Acad Sci U S A* 91:2021–2025.

Pearson CE, Zorbas H, Price GB, Zannis-Hadjopoulos M (1996) Inverted repeats, stem-loops, and cruciforms: Significance for initiation of DNA replication. *J Cell Biochem* 63:1–22.

Potaman VN, Bissler JJ, Hashem VI, Oussatcheva EA, Lu L, Shlyakhtenko LS, Lyubchenko YL, Matsuura T, Ashizawa T, Leffak M, Benham CJ, Sinden RR (2003) Unpaired structures in SCA10 (ATTCT)n.(AGAAT)n repeats. *J Mol Biol* 326:1095–1111.

Potaman VN, Lushnikov AY, Sinden RR, Lyubchenko YL (2002) Site-specific labeling of supercoiled DNA at the A+T rich sequences. *Biochemistry* 41:13198–13206.

Riener CK, Stroh CM, Ebner A, Gall AA, Klampfl C, Romanin C, Lyubchenko YL, Hinterdorfer P, Gruber HJ (2003) A simple test system for single molecule recognition force microscopy. *Analytica Chimica Acta* 479:59–75.

Rivetti C, Codeluppi S, Dieci G, Bustamante C (2003) Visualizing RNA extrusion and DNA wrapping in transcription elongation complexes of bacterial and eukaryotic RNA polymerases. *J Mol Biol* 326:1413–1426.

Rivetti C, Guthold M, Bustamante C (1996) Scanning force microscopy of DNA deposited onto mica: Equilibration versus kinetic trapping studied by statistical polymer chain analysis. *J Mol Biol* 264:919–932.

Rivetti C, Guthold M, Bustamante C (1999) Wrapping of DNA around the E.coli RNA polymerase open promoter complex. *Embo J* 18:4464–4475.

Rivetti C, Walker C, Bustamante C (1998) Polymer chain statistics and conformational analysis of DNA molecules with bends or sections of different flexibility. *J Mol Biol* 280:41–59.

Rode S, Oyabu N, Kobayashi K, Yamada H, Kuhnle A (2009) True atomic-resolution imaging of (1014) calcite in aqueous solution by frequency modulation atomic force microscopy. *Langmuir* 25:2850–2853.

Sacho EJ, Kadyrov FA, Modrich P, Kunkel TA, Erie DA (2008) Direct visualization of asymmetric adenine-nucleotide-induced conformational changes in MutL alpha. *Mol Cell* 29:112–121.

Sakamoto T, Webb MR, Forgacs E, White HD, Sellers JR (2008) Direct observation of the mechanochemical coupling in myosin Va during processive movement. *Nature* 455:128–132.

Santos S, Barcons V, Christenson HK, Billingsley DJ, Bonass WA, Font J, Thomson NH (2013) Stability, resolution, and ultra-low wear amplitude modulation atomic force microscopy of DNA: Small amplitude small set-point imaging. *Appl Phys Lett* 103:063702.

Sarid D (2007) *Exploring scanning probe microscopy with mathematica. Weinheim: Wiley-VGH Verlag GMBH & Co kGaa.*

Scheuring S, Dufrene YF (2010) Atomic force microscopy: Probing the spatial organization, interactions and elasticity of microbial cell envelopes at molecular resolution. *Mol Microbiol* 75:1327–1336.

Shlyakhtenko LS, Gall AA, Filonov A, Cerovac Z, Lushnikov A, Lyubchenko YL (2003a) Silatrane-based surface chemistry for immobilization of DNA, protein–DNA complexes and other biological materials. *Ultramicroscopy* 97:279–287.

Shlyakhtenko LS, Hsieh P, Grigoriev M, Potaman VN, Sinden RR, Lyubchenko YL (2000a) A cruciform structural transition provides a molecular switch for chromosome structure and dynamics. *J Mol Biol* 296:1169–1173.

Shlyakhtenko LS, Lushnikov AY, Lyubchenko YL (2009a) Dynamics of nucleosomes revealed by time-lapse atomic force microscopy. *Biochemistry* 48:7842–7848.

Shlyakhtenko LS, Lushnikov AY, Lyubchenko YL (2009b) Dynamics of nucleosomes revealed by time-lapse atomic force microscopy. *Biochemistry* 48:7842–7848.

Shlyakhtenko LS, Lushnikov AY, Miyagi A, Li M, Harris RS, Lyubchenko YL (2012) Nanoscale structure and dynamics of ABOBEC3G complexes with single-stranded DNA. *Biochemistry* 51:6432–6440.

Shlyakhtenko LS, Lushnikov AY, Miyagi A, Li M, Harris RS, Lyubchenko YL (2013) Atomic force microscopy studies of APOBEC3G oligomerization and dynamics. *J Struct Biol* 184:217–225.

Shlyakhtenko LS, Miloseska L, Potaman VN, Sinden RR, Lyubchenko YL (2003b) Intersegmental interactions in supercoiled DNA: Atomic force microscope study. *Ultramicroscopy* 97:263–270.

Shlyakhtenko LS, Hsieh P, Grigoriev M, Potaman VN, Sinden RR, Lyubchenko Y L (2000a) A cruciform structural transition provides a molecular switch for chromosome structure and dynamics. *J Mol Biol* 296:1169–1173.

Shlyakhtenko LS, Potaman VN, Sinden RR, Gall AA, Lyubchenko YL (2000b) Structure and dynamics of three-way DNA junctions: Atomic force microscopy studies. *Nucleic Acids Res* 28:3472–3477.

Shlyakhtenko LS, Potaman VN, Sinden RR, Lyubchenko YL (1998) Structure and dynamics of supercoil-stabilized DNA cruciforms. *J Mol Biol* 280:61–72.

Sigal N, Alberts B (1972) Genetic recombination: The nature of a crossed strand-exchange between two homologous DNA molecules. *J Mol Biol* 71:789–793.

Sinden RR (1994) *DNA structure and function*. San Diego, CA: Academic Press.

Singh S, Turina P, Bustamante CJ, Keller DJ, Capaldi R (1996) Topographical structure of membrane-bound Escherichia coli F1F0 ATP synthase in aqueous buffer. *FEBS Lett* 397:30–34.

Singleton MR, Scaife S, Wigley DB (2001) Structural analysis of DNA replication fork reversal by RecG. *Cell* 107:79–89.

Sokolov IY, Henderson GS, Wicks FJ (1999) Theoretical and experimental evidence for "true" atomic resolution under non-vacuum conditions. *J Appl Phys* 86:5537–5540.

Sun Z, Tan HY, Bianco PR, Lyubchenko YL (2015) Remodeling of RecG helicase at the DNA replication fork by SSB protein. *Sci Rep* 5:9625.

Suzuki Y, Higuchi Y, Hizume K, Yokokawa M, Yoshimura SH, Yoshikawa K, Takeyasu K (2010) Molecular dynamics of DNA and nucleosomes in solution studied by fast-scanning atomic force microscopy. *Ultramicroscopy* 110:682–688.

Swanson PC (2004) The bounty of RAGs: Recombination signal complexes and reaction outcomes. *Immunol Rev* 200:90–114.

Tessmer I, Yang Y, Zhai J, Du C, Hsieh P, Hingorani MM, Erie DA (2008) Mechanism of MutS searching for DNA mismatches and signaling repair. *J Biol Chem* 283:36646–36654.

Thundat T, Allison DP, Warmack RJ, Brown GM, Jacobson KB, Schrick JJ, Ferrell TL (1992) Atomic force microscopy of DNA on mica and chemically modified mica. *Scanning Microsc* 6:911–918.

Tims HS, Widom J (2007) Stopped-flow fluorescence resonance energy transfer for analysis of nucleosome dynamics. *Methods* 41:296–303.

Tiner WJ Sr, Potaman VN, Sinden RR, Lyubchenko YL (2001) The structure of intramolecular triplex DNA: Atomic force microscopy study. *J Mol Biol* 314:353–357.

Tong Z, Mikheikin A, Krasnoslobodtsev A, Lv Z, Lyubchenko YL (2013) Novel polymer linkers for single molecule AFM force spectroscopy. *Methods* 60:161–168.

Vanamee ES, Viadiu H, Kucera R, Dorner L, Picone S, Schildkraut I, Aggarwal AK (2005) A view of consecutive binding events from structures of tetrameric endonuclease SfiI bound to DNA. *Embo J* 24:4198–4208.

Vesenka J, Guthold M, Tang CL, Keller D, Delaine E, Bustamante C (1992) Substrate preparation for reliable imaging of DNA molecules with the scanning force microscope. *Ultramicroscopy* 42–44(Pt B):1243–1249.

Vohrer U, Hegemann D, Oehr C (2003) XPS, AES, and AFM as tools for study of optimized plasma functionalization. *Anal Bioanal Chem* 375:929–934.

Volcker N, Klee D, Hocker H, Langefeld S (2001) Functionalization of silicone rubber for the covalent immobilization of fibronectin. *J Mater Sci Mater Med* 12:111–119.

Vologodskii AV (2015) *Biophysics of DNA*. Cambridge University Press. Cambridge, UK.

Walters DA, Cleveland JP, Thomson NH, Hansma PK, Wendman MA, Gurley G, Elings V (1996) Short cantilevers for atomic force microscopy. *Rev Sci Instrum* 67:3583–3590.

Wang H, Yang Y, Schofield MJ, Du C, Fridman Y, Lee SD, Larson ED, Drummond JT, Alani E, Hsieh P, Erie DA (2003) DNA bending and unbending by MutS govern mismatch recognition and specificity. *Proc Natl Acad Sci U S A* 100:14822–14827.

Weil R, Vinograd J (1963) The cyclic helix and cyclic coil forms of polyoma viral DNA. *Proc Natl Acad Sci U S A* 50:730–738.

Weisenhorn AL, Egger M, Ohnersorge F, Gould SAC, Heyn SP, Hansma HG, Sinsheimer RL, Gaub HE, Hansma PK (1991) Molecular-resolution images of Langmuir-Blodgett films and DNA by atomic force microscopy. *Langmuir* 7:8–12.

Weisenhorn AL, Römer DU, Lorenzi GP (1992) An atomic force microscope study of Langmuir-Blodgett films of a $\beta^{4.4}$-helical pentadecavaline. *Langmuir* 8:3145–3149.

Wingren R, Elwing H, Erlandsson R, Welin S, Lunström I (1991) Structure of adsorbed fibrinogen obtained by scanning force microscopy. *FEBS Lett* 280:225–228.

Yang J, Mou J, Shao Z (1994) Molecular resolution atomic force microscopy of soluble proteins in solution. *Biochim Biophys Acta* 1199:105–114.

Yang J, Tamm LK, Tillack TW, Shao Z (1993) New approach for atomic force microscopy of membrane proteins. The imaging of cholera toxin. *J Mol Biol* 229:286–290.

Yang Y, Sass LE, Du C, Hsieh P, Erie DA (2005) Determination of protein–DNA binding constants and specificities from statistical analyses of single molecules: MutS-DNA interactions. *Nucleic Acids Res* 33:4322–4334.

Yodh JG, Woodbury N, Shlyakhtenko LS, Lyubchenko YL, Lohr D (2002) Mapping nucleosome locations on the 208–12 by AFM provides clear evidence for cooperativity in array occupation. *Biochemistry* 41:3565–3574.

Zaychikov E, Denissova L, Heumann H (1995) Translocation of the Escherichia coli transcription complex observed in the registers 11 to 20: "jumping" of RNA polymerase and asymmetric expansion and contraction of the "transcription bubble". *Proc Natl Acad Sci U S A* 92:1739–1743.

Zhang Y, Sheng S, Shao Z (1996) Imaging biological structures with the cryo atomic force microscope. *Biophys J* 71:2168–2176.

Zhong Q, Inniss D, Kjoller K, Elings VB (1993) Fractured polymer/silica fiber surface studied by tapping mode atomic force microscopy. *Surf Sci Lett* 290 L688–L692.

3 Atomic Force Microscope Force Spectroscopy

Eric A. Josephs, Piotr E. Marszalek, and Zackary N. Scholl
Duke University
Durham, North Carolina

Contents

3.1 Mechanical Manipulation of Single Molecules with an Atomic Force Microscope

The ability to pull on or stretch individual molecules with subnanometer-scale precision provides a way to directly investigate the underlying physical properties of the biological molecules involved in life's processes. This ability is realized when performing what are called "force spectroscopy" experiments (Figure 3.1), which can be used to reveal the structural and mechanical properties of those biological molecules. During these experiments, how much force is required to extend or deform an individual molecule to a specific length (Figure 3.1b) is recorded while that molecule is being mechanically stretched by an instrument such as an atomic force microscope (Figure 3.1a) or an "optical tweezer" apparatus. These measurements can be thought of as qualitatively similar to "stress-strain" measurements of macroscopic materials by tensometer, brought down to the level of single molecules. Very sensitive

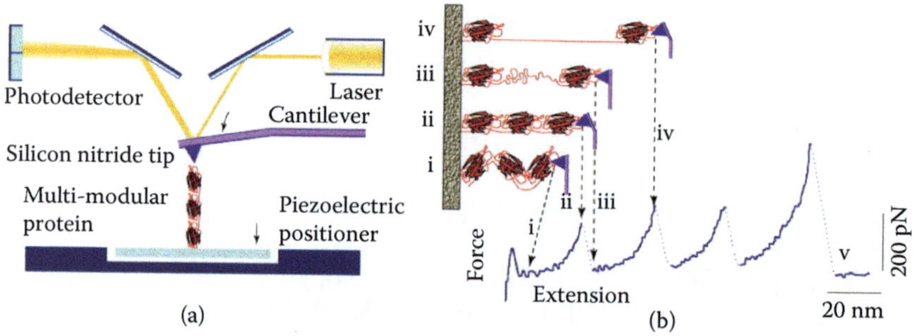

Figure 3.1 (a) AFM performing a force spectroscopy experiment (the unfolding of a multimodular protein). (b) During an unfolding experiment, the tip initially engaged a protein on the surface (i), then as the tip is retracted the protein is mechanically stretched (ii). As the protein is extended, structural components of the protein (domains) will unfold or behave like an unstructured polypeptide chain, producing a sudden drop in the deflection or forces recorded on the tip as the apparent length of the protein increases (iii). The force at which this event occurs is the unfolding force. This process proceeds domain by domain (iv) until the protein detaches from the tip or surface (v). From the force–extension curves produced from these experiments, structural and energetic properties of the protein can be determined. (Adapted from *Trends Biochem. Sci.*, 24, Fisher, T.E., et al., 379–384, 1999. Copyright (1999), with permission from Elsevier.)

measurements of the "force versus extension" characteristics of biomolecules subjected to forces smaller than ~100 pN can be performed with an optical tweezer apparatus (Chapter 5), but force spectroscopy using an atomic force microscope (AFM) has a functional range of approximately 10–10,000 pN, making it ideal, for example, to characterize the forces that hold biological molecules in their functional three-dimensional (3D) conformations; to identify internal molecular transitions that occur when a biomolecule is subjected to moderate or (biologically) large forces; or to determine the mechanical strength of binding or interactions between two biological molecules, all of which can occur across this range of forces. These experiments can reveal structural and physical details about biomolecules, which are largely hidden from other methods of biochemical characterization (Fisher et al. 2000; Sotomayor and Schulten 2007; Muller 2008; Oberhauser and Carrion-Vazquez 2008; Alegre-Cebollada et al. 2010; Noy 2011; Marszalek and Dufrene 2012; Puchner and Gaub 2012; Scholl et al. 2013a; Schuler and Hofmann 2013).

As described in Chapter 2, an AFM uses a piezoelectric actuator to move a mechanical force sensor—a reflective cantilever that has, at its end, a sharp tip with radius of curvature on the order of tens of nanometers—in relation to a biological sample (Figure 3.1a). For a force spectroscopy experiment, the deflections of the cantilevers are used to directly determine the forces of interaction between the AFM tip and the individual molecules that are being mechanically probed by the instrument. This is in distinct contrast to the operation of an AFM in its imaging modes (Chapter 2), where the deflections of the cantilever as it is moved across a biological sample are used to generate a topographical image of it while minimizing the interactive forces between the tip and the sample in order to preserve its integrity.

The ability of an AFM to apply large forces to individual biomolecules allowed for early experiments in the mid- to late-1990s, which recorded the forces needed to stretch DNA (Clausen-Schaumann et al. 1999) and shear individual DNA strands from a DNA double helix (Lee et al. 1994) (Figure 3.2c); pull apart a protein avidin and the small-molecule biotin, which together form one of the strongest biologically occurring noncovalent interactions (Ludwig et al. 1994) (Figure 3.2b); and induce conformational transitions of individual sugar monomers in polysaccharides subjected to mechanical distortion (Rief et al. 1997b; Li et al. 1998; Marszalek et al. 1998; Marszalek et al. 1999; Marszalek et al. 2001; Lee et al. 2012). Another early series of AFM single molecule force spectroscopy (SMFS) experiments had been used to characterize the strong intramolecular interactions that held together the Ig and Fn domains in the muscle protein titin and allowed the protein to reversibly reassume its 3D structure after being mechanically disrupted (Rief et al. 1997a; Kellermayer et al. 1998; Li et al. 2002; Williams et al. 2003). In particular, it was realized that AFM is very well suited to study the biophysical properties of "mechanical" proteins like the muscle protein titin, which serve structural/mechanical roles *in vivo* or those that exhibit significant structural transitions in response to force, which are essential to their functional behavior (Puchner and Gaub 2012).

Additionally, AFM can be used to directly probe the structural characteristics of a wide variety of proteins—even those that do not serve mechanical functions like titin—at the single molecule level (Figure 3.2d), providing a wealth of information

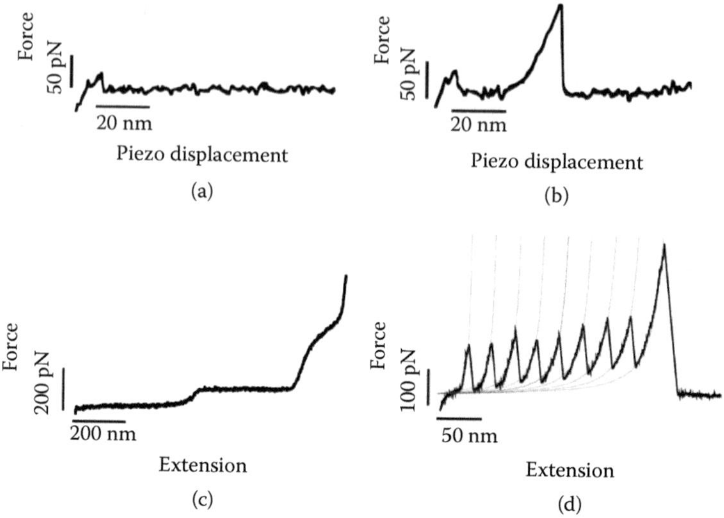

Figure 3.2 Examples of the traces recorded by the AFM tip deflection versus AFM piezo displacements from force spectroscopy experiments with different biological molecules. From the deflection of the cantilever during these experiments, the mechanical properties of biomolecules or the mechanical strength of their interactions can be experimentally determined. (a) An example of a deflection versus piezo displacement curve resulting from nonspecific interactions with a surface. (b) The mechanical rupture of the binding between a single carbohydrate molecule and a lectin (a carbohydrate-binding protein), an example of a force spectroscopy experiment to measure the strength of intermolecular interactions between two biological molecules (see Section 3.3.3). (c) The mechanical stretching of a single molecule of a segment of double-stranded λ DNA molecule exhibiting one structural transition at ~65 pN and a second transition at ~300–350 pN. (d) The sequential unfolding of the I27 domains of the protein titin in an engineered "polyprotein" (i.e., see Figure 3.1b), which can provide insights into the relative stabilities of its component structural elements. Thin red lines are fits of each "unfolding event" to the WLC model of polymer mechanics (see Section 3.2), which can be used to determine the properties of the domain that unfolded. Polyproteins can be engineered at the genetic level to include additional domains from unrelated proteins that are of interest. ([a and b] Adapted by permission from Macmillan Publishers Ltd., *Nat. Meth.*, Hinterdorfer, P., and Dufrene, Y.F., 3, 347–355, 2006. Copyright © 2006. [c] Adapted with permission from Scholl, Z.N., Rabbi, M., Lee, D., Manson, L., S-Gracz, H., and Marszalek, P.E., *Phys. Rev. Lett.*, 111, 188302, 2013. Copyright © 2013 by the American Physical Society with permission.)

about the behavior and biophysical properties of a protein that would otherwise be very difficult to resolve using more traditional biochemical techniques. For example, AFM force spectroscopy can be used to detect short-lived or heterogeneous interactions within the proteins (Zhuang and Rief 2003); to record the effects of ligand binding or oligomerization on the stability of specific parts of a protein (Hanley et al. 2003; Wojcikiewicz et al. 2006; Bertz and Rief 2009; Scholl et al. 2015); to examine protein unfolding and refolding pathways and mechanisms (Carrion-Vazquez et al. 1999; Fernandez and Li 2004; Bornschlögl and Rief 2006; Lee et al. 2006; Junker et al. 2009; Lee et al. 2010; Stigler et al. 2011; Stigler and Rief 2012; Scholl et al. 2014); and to extract individual membrane proteins from biological membranes (Muller et al. 2000; Oesterhelt et al. 2000) as well as to monitor the insertion of these proteins back into those membranes (Kessler et al. 2006). AFM SMFS can also be used to directly measure the strength of interactions between two biological molecules (Kienberger et al. 2006; Kim et al. 2011), such as between two proteins or an antibody and its antigen, making SMFS a very versatile technique (Figure 3.2b). However, compared with other spectroscopic techniques, which record the average structural properties of a large number of biomolecules in a sample simultaneously—and, hence, often lose these abovementioned benefits—the practical limitation for AFM is its throughput, as a single protein may take many experiments until enough individual molecules are probed so the protein is fully characterized.

In this chapter, we describe the principal aspects, advantages, and applications of force spectroscopy using an AFM, paying particular attention to the technical differences between the its imaging functionalities and those for force spectroscopy. In Sections 3.2 and 3.3, we illustrate the practical considerations and protocols necessary to perform and analyze an AFM force spectroscopy experiment. In Section 3.4, we provide recent examples from the literature demonstrating the utility of AFM SMFS in answering poignant biological questions that would be very difficult to approach otherwise. We should note as well that the specific practical aspects of molecular recognition force spectroscopy—which uses an AFM to measure and map the binding interactions between isolated or membrane-bound receptor–ligand molecular partners, in particular on the surfaces of living cells—is not described in great detail here (except as in Section 3.2.2); the reader is referred to several excellent reviews on this subject (Dufrêne and Hinterdorfer 2008; Muller et al. 2009; Willet et al. 2012).

3.2 Principles of Conducting a Force Spectroscopy Experiment

While force spectroscopy using an AFM has been widely used to examine all major biomolecules such as nucleic acids, sugars (carbohydrates), and proteins (Marszalek and Dufrene 2012) (Figure 3.2), in this section we illustrate the principles of conducting a typical force spectroscopy experiment with an AFM using the mechanical denaturing or "unfolding" of a protein as an example (such as in Figure 3.1). Proteins are large biomolecules composed of a chain of different amino acids that are specific to that type of protein. They are biologically important for a variety of reasons: in particular, they serve mechanical or structural roles within cells, can actively transport other molecules around within a cell, and are responsible for catalyzing the majority of all chemical and biological reactions required for life. For the protein to perform these functions properly, the amino acid chain must first "fold" into a unique, 3D conformation that is determined by its chemical composition and sequence (the order of amino acids). Understanding the intramolecular forces that hold the protein in these functional conformations and understanding how the amino acid sequence of a protein is able obtain this 3D structure is critical for many applications of molecular biology, drug design, and etiology.

In this endeavor, it is often useful to perturb the proteins, frequently achieved using heat or chemical denaturants, to determine the effects of these disruptive agents and to establish the stabilities of different structural features of the protein. By directly interacting with individual protein molecules, an AFM can "mechanically" perturb these structures and can be used to measure the structural and mechanical properties of these features of a protein directly. This ability of an AFM to manipulate individual molecules has an added benefit of alleviating protein aggregation, which often occurs in solutions of proteins whose structures have become highly perturbed (denatured).

During a typical experiment, an AFM tip is brought into contact with a surface onto which a biomolecule of interest has been deposited, and then retracted in order to mechanically stretch one of those molecules that may have adsorbed onto it during the time the tip was on the surface. From the deflection of the cantilever during this process, both the mechanical properties of the biomolecule and, of particular interest, the forces required to mechanically denature all or parts of a protein can be experimentally determined. In this chapter, we provide an overview of and considerations to be made before beginning a typical SMFS experiment using an AFM.

3.2.1 Considerations Prior to an AFM Single Molecule Force Spectroscopy Experiment

In principle, SMFS measurements can be carried out on many commercial or home-made AFM instruments (Oberhauser et al. 1998; Rabbi and Marszalek 2008). However, the requirements for performing an SMFS experiment using an AFM are different than those for performing imaging with an AFM, so AFM instruments used for SMFS should be optimized specifically for these applications. Additionally, the sample and equipment are prepared differently prior to a force spectroscopy experiment compared to when an AFM imaging experiment is to be performed. These modifications to the equipment and preparatory procedures are related to maintaining the sensitivity of an instrument required to measure mechanical displacements and forces at biologically relevant scales and its capacity to unambiguously identify single molecules.

In general, more precise control of the piezoelectric manipulator requires the "closed-loop" control of the feedback in the direction of pulling. During closed-loop control, high-precision position sensors actively monitor the piezo, and the outputs of these sensors are used by the piezoelectric actuators to actively perform high-speed adjustments to the position of the piezo in order to correct for instrumental errors (such as hysteresis or drifts) that may occur. Additionally, in contrast with those stiff AFM cantilever tips used for AFM imaging, soft cantilevers (having spring constants of ~5–20 pN/nm) are best for force spectroscopy that typically involves forces between 10 and 1000 pN. Stiffer cantilevers can be used to measure forces from 100 to 10,000 pN. The spring constant k of a rectangular cantilever can be approximated by its geometric properties using solid-beam mechanics:

$$k = \frac{E_c w t^3}{4L^3}$$

where E_c is the elastic modulus of the cantilever material; and w, t, and L are the width, thickness, and length of the cantilever, respectively (Butt et al. 2005). Hence, a cantilever is designed to be soft by increasing the length of the cantilever and decreasing the thickness. However, since the resonance frequency of the cantilever is proportional to (t/L^2), it is best to decrease the length so that the resonant frequency can be as high as possible, since use of cantilevers with high-resonance frequencies makes it possible to effectively filter the thermal fluctuations of the cantilever (thermal noise) from the measurements. Thus, a cantilever is usually designed to be very thin and very short for an optimal spring constant.

Using ultra small cantilevers, such as BioLevers (Olympus or Bruker), with the spring constant of ~5 pN/nm, it is possible to resolve forces as small as 5–10 pN when the measurement bandwidth is limited to around 1 kHz. Of note is the fact that by lowering the measurement bandwidth by filtering the photodiode signal with a low-pass electronic filter, it is possible to eliminate some thermal noise of the cantilever and improve force resolution at the cost of missing some very fast events. Application of more complex methodologies such as lock-in force spectroscopy (Schlierf et al. 2007) makes it possible to further improve force resolution to ~1 pN.

Before the SMFS experiments can begin, the tip-holder/cantilever apparatus should be immersed in the liquid solution in which the experiment is to be performed for approximately 30 minutes to an hour so that it may become thermally equilibrated. Otherwise, different thermal coefficients of expansion between the cantilever material (generally silicon or silicon nitride) and reflective coatings applied to the backside of the cantilever (generally gold) will result in large deflections (drift) in the z-direction as the cantilever comes to thermal equilibrium with its surroundings. Another approach to improve force resolution and minimize this drift is to chemically strip the reflective gold coating from the AFM cantilever (Sullan et al. 2013); however, in the absence of a reflective coating, this modification typically requires the use of infrared lasers to measure cantilever deflection, which not all AFM instruments support.

Precise determination of the cantilever spring constant is necessary for the accurate determination of force being applied to a biomolecule. As it is, the spring constant of an AFM cantilever can vary up to 10%–20% from its nominal or manufacturer-reported average value (and whose value is less vital to know for AFM imaging experiments), so the spring constant of each cantilever must be determined immediately prior to each experiment. Two methods are generally used to determine the spring constant of an AFM cantilever: the "geometric" method discussed previously, where the mechanical properties of cantilevers are estimated from measurement of the cantilever shape using an optical microscope; and the "thermal noise" or the "energy equipartition" method, where the spring constant is derived from thermal fluctuations of the free cantilever itself (Florin et al. 1995; Butt et al. 2005). In practice, we prefer the thermal noise method, as it can be performed simply and immediately prior to starting the experiment within the experimental setup (Box 3.1).

Finally, considerations must be made as to the selection of the protein to be manipulated. In general, as discussed in Section 3.3, knowing the 3D structure

Box 3.1 Calibration of the AFM Cantilever Spring Constant Using the Thermal Noise Method

It is critical that prior to each experiment the spring constant for the AFM cantilever to be used is determined *in situ*. To determine the spring constant k of an AFM cantilever using the thermal noise method (Florin et al. 1994; Florin et al. 1995; Butt et al. 2005), the tip is immersed in the environment under which the experiment will be performed (e.g., protein buffer solution) and the fluctuations of the free cantilever far away from surface are recorded as a raw deflection signal. The cantilever is assumed to behave as a Hookean spring, and according to the equipartition theorem:

$$k = \frac{k_B T}{s^2 \langle z^2 \rangle}$$

where k_B is Boltzmann's constant, T is the absolute ambient temperature, $\langle z^2 \rangle$ is the mean-squared deflection of the cantilever over time, and s is the sensitivity (the experiment-specific conversion factor between actual cantilever deflection and the recorded deflection signal from the photodiode). In practice, rather than attempting to calculate the mean-squared deflection of the cantilever $\langle z^2 \rangle$ from the time course of the deflection signal, it is calculated by numerically integrating over the first mode of the power spectrum of the cantilever fluctuations, which itself is found by squaring the Fourier transform of the deflection signal (Figure 3.3a). Observing the cantilever thermal fluctuation in the "frequency" rather than "time" domain allows evaluation of other, nonthermal oscillations of the cantilever (e.g., due to mechanical noise from the environment) that may be significant at low frequencies. Noise from the environment should be minimized by using vibration isolation platforms such as mounting the AFM on an air table and enclosing it within an acoustic chamber, both of which are common practices in AFM imaging applications as well.

At room temperature, $k_B T$ is about 4.1×10^{-21} J or 4.1 pN · nm, and $\langle z^2 \rangle$ will be in units of V^2 from the deflection signal measured by position-sensitive photodiode of the AFM. To determine the conversion between deflection by the actual cantilever and the recorded deflection signal in V, the piezo is engaged and contact is made between the cantilever and a *hard* surface (Figure 3.3b). The slope of the contact in the cantilever deflection–piezo displacement curve yields the sensitivity s, generally in units of Å/V or nm/V, and the spring constant k can be determined.

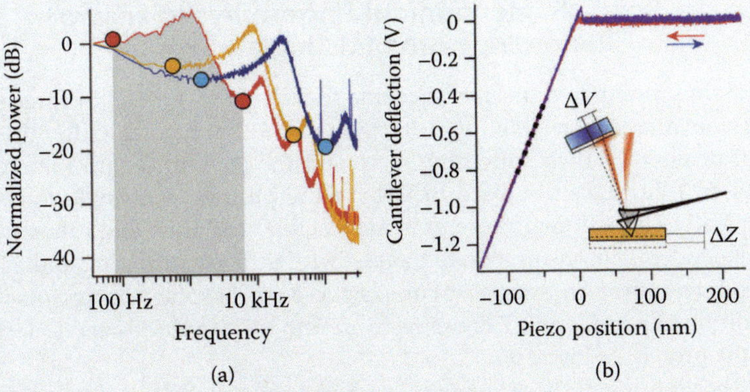

Figure 3.3 (a) By integrating over the first mode of the power spectrum of the thermal cantilever fluctuations (boundaries of integration marked by the two circles for each of three cantilevers of increasing stiffness—red, yellow, and blue), the mean-squared deflection of the cantilever $<z^2>$ is found. (b) The sensitivity *s* of the cantilever is ascertained from the slope of the cantilever deflections versus piezo extension signal upon contact with the surface (dotted line). The spring constant of the cantilever is then determined according to the equipartition theorem. (Adapted by permission from Macmillan Publishers Ltd., *Nat. Protoc.*, Popa, I., et al., 8, 1261–1276, 2013. Copyright © 2013.)

from X-ray crystallography or nuclear magnetic resonance (NMR) will help in the interpretation of the force versus extension curve (*F*–*X* curve), the product of a SMFS experiment. Additionally, certain modifications to the protein at the molecular level can be performed, which in general do not significantly alter protein structure or function but will aid in single molecule manipulation and analysis. Some of these modifications can be incorporated at the genetic level prior to purifying the protein for experimentation. For example, the inclusion of nonnative cysteine residues at one terminus of the protein can help in orienting the molecule onto a gold surface through the strong gold–thiol interaction, as can the inclusion of known peptide sequences, which will bind tightly or covalently to surfaces that have been prefunctionalized with their specific antigen (such as HaloTag [Taniguchi and Kawakami 2010], e.g., or monovalent Strep-Tactin [Baumann et al. 2015]). Also, incorporating, at the genetic level, a secondary protein with a known *F*–*X* curve at both ends of the protein of interest (in what is known as a "polyprotein" construct) can also help to ensure the experimenter that he or she is manipulating a single molecule by providing a molecular "fingerprint" in the *F*–*X* curve; see Box 3.2 for extended discussion of the use of these molecular fingerprints in SMFS. Examples of these incorporated features with well-characterized fingerprints include I27 polyproteins (Carrion-Vazquez et al. 1999; Perez-Jimenez et al. 2006; Dougan and Fernandez 2007), GB1 domains (Cao and Li 2007), and filamin domains (Schwaiger et al. 2004; Bornschlögl and Rief 2006).

Box 3.2 Mechanical Fingerprints to Ensure Recordings Are of a Single Molecule

Certain protein domains have become so well characterized by force spectroscopy that, when added at the genetic level either before or after the protein of interest, their unfolding force and the contour length increments in a *F–X* curve can be used to test for the pulling of a single molecule. Typically, these domains are extremely stable, and they fold robustly and independently of the protein of interest with little perturbation to that protein's structure. An example of one of these mechanical fingerprints is the addition of a series of I27 domains from titin to both the N- and C-termini of the protein of interest.

The addition of these domains can help to ensure that only a single molecule is being manipulated at a time, and to ensure forces applied are being applied across the entire protein of interest. For example, consider the addition of two I27 domains both before and after the protein of interest. I27 domains are quite stable and will unfold independently at around 200 pN at typical pulling loading rates with a contour increment of 28 nm (see, e.g., Figure 3.5). If during an SMFS experiment, at least three unfolding events with these characteristic unfolding forces and contour increments are observed in a *F–X* curve, we can be reasonably certain that the other unfolding events have been caused by forces applied across the entire protein of interest, as there are only two I27 domains on each side. Additionally, if we observe more than four I27 domains, or if the unfolding forces of the I27 domain appear significantly low (perhaps half the expected force), this may imply that more than one molecule is being manipulated during an extension. Typically, when first interrogating a new protein, it is best to identify the *F–X* curves that contain all four I27 in a full extension, so that the mechanical fingerprint of the protein of interest can be determined.

A number of these domains have been characterized and can be selected based on the demands of the specific experiment, such that the forces of those domains do not significantly overlap with those of the protein of interest. For example, filamin domains (Schwaiger et al. 2004; Bornschlögl and Rief 2006) as well as GB1 domains (Cao and Li 2007) have been used.

3.2.2 Performance of an SMFS Experiment Using an AFM

To prepare the sample for an AFM SMFS experiment, a glass, mica, or gold-coated sample substrate is first stringently cleaned by rinsing with solvent, then dried. For some experiments, the surface can be chemically modified to present different chemical functional groups to the surface in order to modify the interactions between the protein and the surface or to present a specific antigen to the protein, as mentioned above. Gold can be modified by immersing the sample in a dilute (~10 μM to 1 mM) solution of thiols (particularly alkanethiols) to form ordered, self-assembled monolayers on their surfaces (Love et al. 2005), while silanes are typically used to covalently modify the surfaces of glass or mica (Ulman 1996) by evaporation under inert gas or direct adsorption from dry, organic solutions (silane modification of glass or mica should be performed under inert gas in an enclosed glovebox). Recently, silatrane chemistries have been developed for the modification of glass or mica substrates (Shlyakhtenko et al. 2003) and have the distinct advantage of being extremely stable under aqueous or ambient conditions, which dramatically simplifies the modification of mica or glass.

A small drop of a solution containing purified molecule of interest—for example, a protein—with approximate concentrations of 0.01–2 mg/mL, is deposited on the surface for approximately a half hour and then gently rinsed with buffer. During this incubation, the proteins become chemically bound to the surface or are adsorbed onto the surface via direct physisorption, and the sample is ready for force spectroscopy.

After the spring constant of the tip k has been calibrated (Box 3.1), the piezo is engaged and the tip is brought into contact with the surface and then retracted at constant speeds (typically between 100 and 1000 nm/s). Most often (~90%–99.9% of the time) the tip does not pick up a protein, although occasionally some spurious, high-force deflection signals may be observed at short distances from the surface (Figure 3.2a). These signals are indicative of contaminations or poor absorption of the protein to the tip that results in nonspecific adhesion of the tip to the surface and can generally be disregarded. However, if the tip does comes into contact with an adsorbed protein, the protein may become simultaneously stuck to both the AFM tip and the surface, and the protein will be mechanically stretched between the two (Figure 3.1i). A change in the deflection signal z can then be used to interpret the forces F being applied to the protein during retraction: $F = -k\,z$. It is also important to remember that the "displacement" d of the piezo does not reflect the position of the cantilever, which itself will bend in response to forces applied to it. The relationship between piezo displacement and the "extension" of the protein, X, can be determined by a simple formula:

$$X = d - F / k - d_0$$

where d_0 is the piezo displacement at the point where the tip first comes into direct contact with the surface. The F–X curve contains the necessary information for investigations of the mechanical properties of individual proteins.

If it is possible for the individual units of a biomolecular polymer to rearrange themselves structurally, they will deform to favor an extended conformation in response to applied forces. As the tip is retracted and mechanical stretching of the protein continues, the deformation of the protein is typically well described by a model of polymer mechanics known as the worm-like chain (WLC) model (Bustamante et al. 1994; Marko and Siggia 1995; Bouchiat et al. 1999) (see, e.g., Figure 3.2d or 3.4a). This model assumes the manner in which the protein will stretch is determined by its total contour length L_c, or the length of polymer backbone (which is about 0.365 nm per amino acid [Dietz and Rief 2006]); as well as its persistence length P, a measure of the stiffness of the polymer (for many proteins or polypeptide chains of interest, P is approximately 0.4–1 nm;

however, for rod-like proteins such as actin, P would be significantly greater, although these proteins are rarely stretched by AFM). That is, the protein requires force F be applied to it in order to stretch to extension X according to an approximate formula:

$$F = \frac{k_B T}{P}\left(\frac{1}{4}\left(1 - X/L_c\right)^{-2} + X/L_c - 1/4\right)$$

Additional details as well as a derivation of the WLC model can be found in Bouchiat et al. (1999). In general, proteins and other biomolecular polymers that deform in the manner of a WLC will stretch approximately linearly (like a spring) at low forces, and then, as they approach their maximum possible length, the forces required to stretch them will increase sharply. Other biomolecular polymers, such as polysaccharides, may be better described by different models of polymer mechanics such as the freely jointed chain model (Flory 1953).

Proteins are composed of structurally well-defined elements known as domains, and the effective contour length of the protein L_c remains relatively small, while the domains of that protein remain folded, or retain a conformation similar to its conformation in the absence of applied forces. As the protein's end-to-end distance begins to approach its effective contour length during stretching, the individual domains within that protein can unfold suddenly as forces applied across that domain surpass the internal forces holding its structure together (Figure 3.1ii). This unfolding will result in a sudden decrease in cantilever deflection as the effective contour length of the protein increases to include the length of an unstructured polypeptide chain (ΔL_c = [the length of the newly unfolded polypeptide chain] – [end-to-end length of the folded domain]). The peak force observed immediately before this unfolding event is observed is known as the unfolding force F_u. This process of unfolding will occur domain by domain as the tip is retracted until the entire protein is extended to an unstructured polypeptide chain (Figure 3.1iii and iv). Once the force applied exceeds the forces holding the protein to the AFM tip (or to the surface), the protein will desorb from the tip (or surface). Upon detachment, the forces applied to, and measured by, the AFM tip will drop to zero (Figure 3.1v).

Because of molecular-scale thermal fluctuations that occur within the protein in solution, it is important to note that the unfolding force F_u will vary stochastically about a mean value for each extension by an AFM. Additionally, the mean unfolding force is determined by the loading rate ν of forces applied to the protein, or the product of the spring constant of the AFM tip and the speed of the piezo retraction (pN/s). As discussed in Section 3.3.1 in more detail, by repeating the series of unfolding experiments at different loading rates, detailed information regarding the thermodynamic properties of protein unfolding can be extracted. These experiments are repeated until about 50 or more F–X curves of the full protein are acquired. Subsequent interpretation of complex F–X curves and the assignment of the unfolding events to the individual domains of the protein can be challenging, although physical models and available computational tools can be helpful in this regard.

Additionally, in an alternative force spectroscopy paradigm, the force applied to a single protein can be held constant over time using high-speed piezo feedback controls in what are termed "force clamp" experiments (Oberhauser et al. 2001; Fernandez and Li 2004). In these experiments, the length of the protein, rather than the applied forces, is observed over time at constant applied forces in "force spectrograms." This setup allows the rates of protein unfolding or refolding at different applied forces to be observed over long periods of time.

3.3 Interpretation of the Force–Extension Curves in SMFS Experiments

Although an individual *F*–*X* curve can be illuminating, it is best to check the consistency of the *F*–*X* curves by superposing multiple *F*–*X* curves and comparing the respective properties of their unfolding peaks. Each individual curve can be analyzed with respect to its *unfolding forces* as well as to the *contour lengths* by fitting each peak to the WLC model (Figure 3.4a).

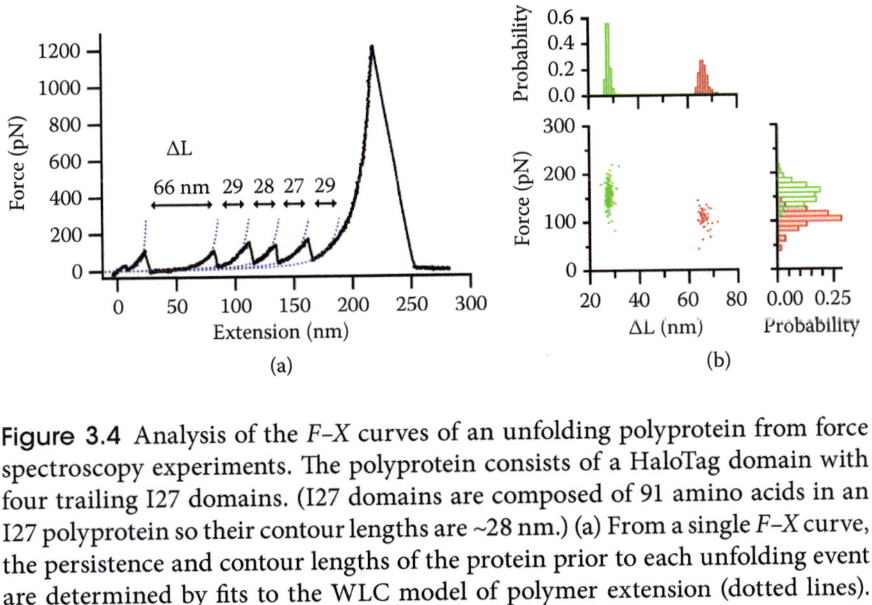

Figure 3.4 Analysis of the *F*–*X* curves of an unfolding polyprotein from force spectroscopy experiments. The polyprotein consists of a HaloTag domain with four trailing I27 domains. (I27 domains are composed of 91 amino acids in an I27 polyprotein so their contour lengths are ~28 nm.) (a) From a single *F*–*X* curve, the persistence and contour lengths of the protein prior to each unfolding event are determined by fits to the WLC model of polymer extension (dotted lines). From the differences in contour lengths (Δ*L*), the number of amino acids in the unfolded domains can be determined. (b) After repeating this analysis multiple times, the distributions of the different unfolding forces and contour length increments can reveal the structural properties of the unfolded domains (those of the I27 domains [green] and of the HaloTag domain [red]) and additional analyses can be performed. (Reprinted with permission from Taniguchi, Y., and Kawakami, M., *Langmuir*, 26, 10433–10436, 2010. Copyright © (2010) American Chemical Society.)

The difference between individual contour lengths gives the contour length increment (ΔL), which can be directly related to the number of amino acids that were involved in the particular unfolding event (~0.37 nm/amino acid residue). Once many curves have been analyzed, the contour length increments and unfolding forces are compiled into a scatterplot that can show the distributions of both measurements (Figure 3.4b). This can be further analyzed to determine whether a protein unfolds as a single domain or in multiple domains, and what the relative strengths of the protein are.

Furthermore, computational tools in various programs, such as NAMD (Phillips et al. 2005) or GROMACS (Pronk et al. 2013), can simulate the mechanical denaturing of proteins through a process called *steered molecular dynamics* (Isralewitz et al. 2000). These simulations allow researchers to model the unfolding of individual domains, their respective mechanical strengths, and the unfolding pathway of the entire protein (or the order in which the domains mechanically denature). A limitation of these simulations at present is that, for detailed simulations that model the behavior of all of the atoms of the protein subjected to applied forces and of the surrounding solution, pulling rates must be very fast (relative to experimental pulling rates) in order for the computations to complete after a practical period of time (in the order of days); otherwise coarse-grain simulations that make significant simplifying assumptions regarding the intramolecular behaviors of the protein must be performed. While the performance of high-speed SMFS experiments can help bridge this gap between computational and experimental studies (e.g., see Box 3.4), direct comparison of these "all-atom" simulations or these "coarse-grain" simulations with AFM SMFS experiments, therefore, can only provide qualitative guidance for the interpretation of *F–X* curves. Recent advances, such as the development of computational all-atom Monte Carlo pulling (MCP) (Zhang and Lyubchenko 2014), are addressing this challenge.

We also note that "coiled coil" motifs can be inserted at the genetic level into specific domains within the protein of interest for subsequent experiments to be used as an additional tool to characterize the unfolding pathway experimentally (Li et al. 2014). These coiled coils have a very short end-to-end distance when folded, minimally perturb the structure of the domain into which they are inserted, and fold independently of that domain. While the coiled coil "probe" is mechanically isolated while that domain remains folded, when that domain unfolds the coiled coil is now subjected to mechanical forces and will register as a new unfolding peak. This technique permits unfolding events to be unambiguously assigned to specific domains if there are multiple unfolding peaks observed in the *F–X* curve with similar contour lengths and unfolding forces.

There is a major caveat when interpreting the SMFS data, however, it is critical to determine whether or not an individual protein, and not multiple proteins, has been mechanically stretched during a single force curve. This cannot necessarily be determined out of hand, even if the extension of the protein prior to detachment from the tip is similar to the expected contour length of the full protein of interest. This needs to be checked through analysis of the data after capturing multiple *F–X* curves (to see, e.g., if the strength of unfolding forces of an individual peak appears multimodal). Unambiguous determination only a single molecule is being manipulated at a time can also be accomplished by using or introducing protein "fingerprints," or additional protein domains or features that produce a consistent characteristic shape in the *F–X* curve, along with the protein of interest (Box 3.2).

3.3.1 Dynamic Force Spectroscopy

After this initial analysis, it is often useful to take distributions of unfolding forces and determine the intrinsic unfolding rate (i.e., the unfolding rate in the absence of force) (Carrion-Vazquez et al. 1999) to compare the results of SMFS experiments with what would be obtained from traditional biochemical experiments. This comparison can be accomplished by performing what are termed *dynamic force spectroscopy* (DFS) experiments. DFS provides a framework to relate the unfolding of proteins under the nonequilibrium conditions (as they are being stretched by a cantilever with a particular loading rate) to the energetic properties of the protein structure at equilibrium (Box 3.3). The procedure is simple: after many successful experiments are performed at different pulling speeds or loading rates (e.g., by using progressively stiffer cantilevers), the distribution of observed unfolding forces of the target protein at each speed can be tabulated in a histogram. The mode

Box 3.3 Modeling Protein Unfolding as a Diffusive Escape Over a Barrier

A protein in its folded state—that is, its most stable and energetically favorable conformation (left well in Figure 3.5)—can be considered as residing at an energetic minimum of its conformational space (ΔG_{bu} being the free energy difference between bound and unbound state). In order for the protein to unfold to a less energetically favorable state, significant intramolecular interactions must be disrupted, which can be considered as an energy barrier ($\Delta G^{\ddagger}$, located at $x_u^{\ddagger}$ in Figure 3.5) that keeps the protein folded in this conformation. It is unlikely that the protein will be able to traverse this barrier to unfold as a result of thermal energy alone (solid black curve) unless the protein is heated to very high temperatures via thermal denaturation or the energy barrier is lowered via chemical denaturation. By applying forces across the protein (dashed red line), the energy landscape of the conformational space is effectively tilted due to the mechanical work, $F \times X$, performed on the protein. This tilting of the energy landscape (dotted red curve) results in the lowering of the energetic barrier, exponentially increasing the rate at which the protein can then spontaneously unfold, while simultaneously lowering the energy of the unfolded state, making it less likely that the protein will spontaneously refold.

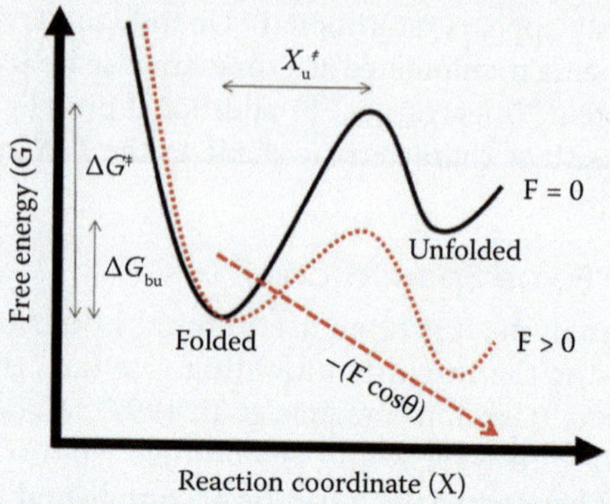

Figure 3.5 Free energy (*G*) landscape of the protein conformational space. Coordinate *x* can be considered the protein end-to-end distance or pulling coordinate during a force spectroscopy experiment. (black curve) In the absence of applied forces, the folded state (left well) is more energetically favorable than its unfolded state (right well). Upon application of forces (dashed red line), the energy landscape is tiled by the mechanical work performed (red dotted line), so that the extended or unfolded state is now more energetically favorable and the energetic barrier between the two states is lowered, exponentially increasing the rate of spontaneous unfolding.

or most likely rupture force—found at the peak in the force histograms—observed for these experiments is then plotted against the respective loading rate for those experiments.

The relationship between loading rate ν and the most likely unfolding force $F_u(\nu)$ (or rupture force, in the case of SMFS experiments measuring ligand–antigen binding strengths) is roughly log-linear (Evans and Ritchie 1997) (Figure 3.6), as approximated using the Bell–Evans equation (Merkel et al. 1999):

$$F_u(\nu) = \mathrm{k_B T}/x^{\ddagger} \log\left(\frac{\nu x^{\ddagger}}{k_u \,\mathrm{k_B T}}\right)$$

where $x^{\ddagger}$ is the distance to the transition state (Box 3.2 and Figure 3.5), and k_u is the intrinsic unfolding rate (or detachment rate, in the case of ligand-antigen AFM SMFS measurements) in the absence of applied forces (Evans et al. 2010). Thus, by fitting these data to the Bell–Evans equation, intrinsic properties of the protein unfolding energy landscape or ligand–receptor binding energy landscape can be determined. The identification of multiple piecewise linear regions of the $F_u(\nu)$ versus log(ν) plots that are each well-fit by different parameters of the Bell–Evans equations (Figure 3.6) has been interpreted

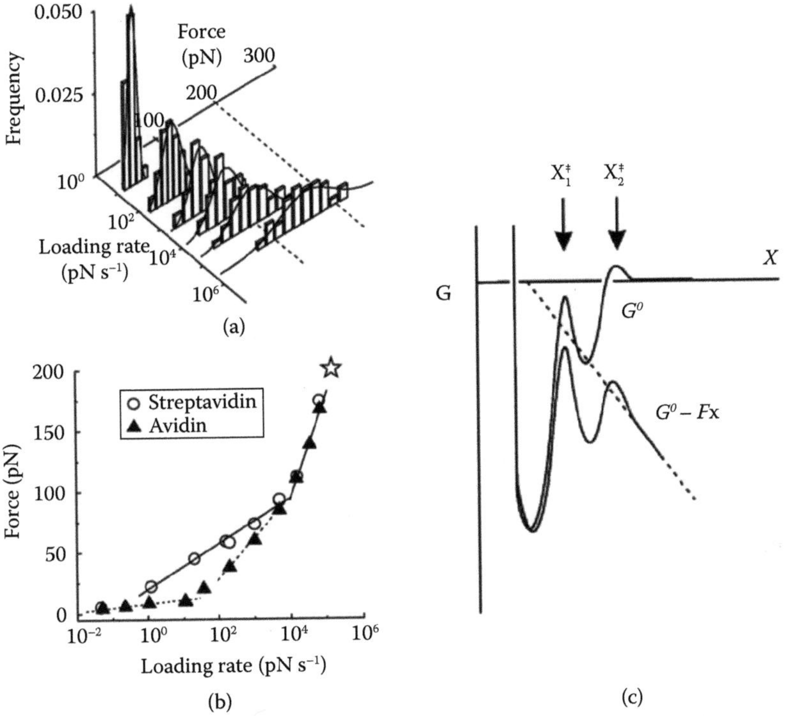

Figure 3.6 Example of DFS analysis of biotin-streptavidin and biotin-avidin (ligand-antigen) interactions. (a) Histograms of the rupture forces for biotin-streptavidin interactions. (b) Most frequent rupture force versus loading rates. The high-strength regime data (star) were collected using AFM SMFS. Note several distinct piecewise linear regimes in the data. (c) A complex energy landscape with a transition state. As the loading rate changes, the energy landscape is tilted by $F \times x$ (note that $x = v \cdot t$, where t is time), so that while the rate of traversal over the energetic barrier at $x^{\ddagger}_2$ initially dominates the overall rupture rate, at higher loading rates traversal over the energetic barrier $x^{\ddagger}_1$ becomes a rate-limiting event. Similar energetic landscapes can be determined for the complex energetic landscapes of protein (un)folding using DFS. (Adapted by permission from Macmillan Publishers Ltd., *Nature,* Merkel, R., et al., 397, 50–53, 1999. Copyright © [1999].)

as resulting from complex energy landscapes with multiple energy minima (termed *transition states*) (Figure 3.6c).

The Bell–Evans equation has been applied quite successfully to the interpretation of *F–X* curves of protein unfolding and ligand–receptor interactions. In typical SMFS experiments performed using an AFM, the loading rates span 10^2–10^5 pN/s, and obtaining the peak unfolding (or rupture) forces at several loading rates within this range is often sufficient to determine energetic properties of interest using DFS. Recent instrumental advances have expanded the limits of possible loading ranges that can be applied experimentally so that new temporal regimes can be explored using AFM force spectroscopy (Box 3.4). More recent models by Szabo, Hummer, and Dudko (Hummer and Szabo 2001; Dudko et al. 2006) as well as by Friddle, Noy, and De Yoreo (Friddle et al. 2012), which extend this foundation by relaxing some of the simplifying assumptions, have been shown to fit experimental data quite well in practice across a wide range of loading rates.

3.3.2 Protein Refolding Experiments

If the protein does not detach from the AFM tip after an approach/retraction cycle—for example, if the distance retracted from the surface is specified to be less than the full contour length of the protein of interest—domains of the protein may spontaneously refold when the forces applied to the protein are relaxed (Rief et al. 1997a; Oberhauser et al. 1998; Carrion-Vazquez et al. 1999). Typically, the tip is quickly brought back to the surface after unfolding in order to minimize any applied forces so that the protein may be able to spontaneously refold without bias. After waiting a specified amount of time (5 milliseconds–10 seconds, typically), the molecule is stretched and relaxed again. The proportion of force peaks seen in the second trace compared to the first *F*–*X* curve can be used to determine the probability that a domain will

Box 3.4 Pushing the Limits of AFM Force Spectroscopy

More recently, improvements to instrumentation and to fabrication of cantilevers have allowed for ultrastable AFM force spectroscopy (Sullan et al. 2013), which allows extremely low loading rates. This allows for measurements of protein folding/unfolding forces in a near-equilibrium regime (He et al. 2015). At the opposite end of the spectrum, extremely high loading rates can be achieved using fast pulling speeds (~4000 μm/s) similar to high-speed AFM imaging (Rico et al. 2013). High loading rates allow for some unstable intermediate states to be identified, which may not be observed at lower loading rates. However, at these speeds hydrodynamic effects on the cantilever must be taken into consideration.

Furthermore, while protein folding events typically can occur on ~1 ms time scales, fast structural fluctuations or short-lived intermediates may also be present, and capturing these events during an AFM SMFS experiment can help to elucidate novel structural features. However, the typical response time of soft commercial cantilevers is on the order of ~500 microseconds. Recently, Edwards et al. demonstrated a technique to modify commercial AFM tips using a focused ion beam (FIB) to produce soft cantilevers (20 pN/nm) with sub-1 microsecond response times that also significantly reduced resonant Brownian oscillations during pulling (Edwards et al. 2015) (Figure 3.7). This and related advances push the limits of the events capable of being recorded by AFM SMFS.

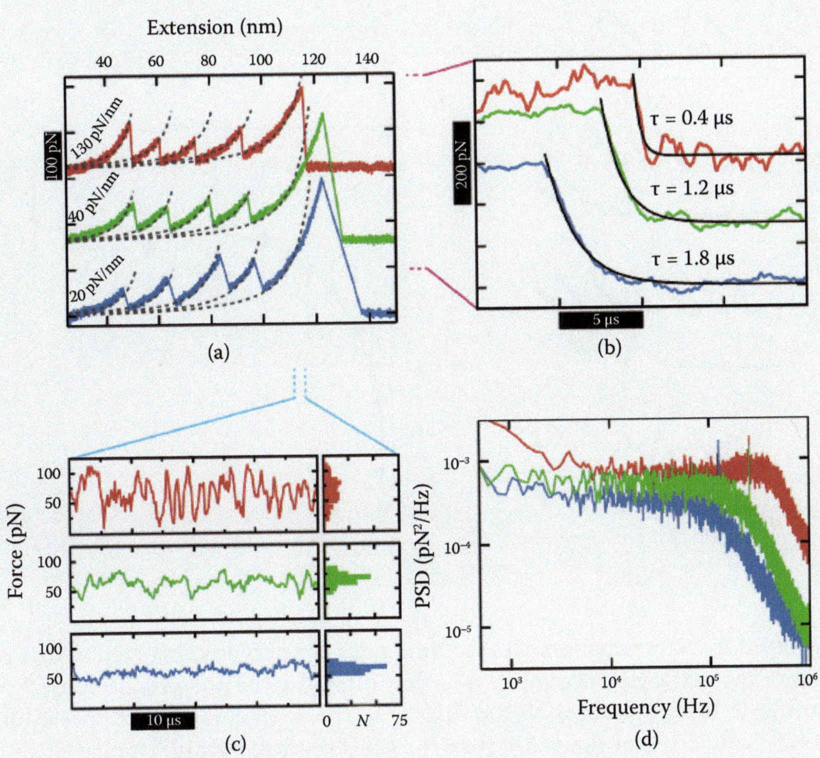

Figure 3.7 Demonstration of ultra-short response times on a commercial AFM using modified cantilevers. (a) Force versus extension curves of an NuG2 polyprotein using a BioLever Fast cantilever (red), a stiff BioLever Fast cantilever modified using a FIB via the method of (Edwards et al. 2015) (green), and a softer BioLever Fast also modified by an FIB (blue). Dashed lines are fit to the WLC model. (b) Force versus time traces of the cantilevers after polyprotein detachment show that the unmodified cantilever (red) exhibits underdamped "ringing," which is not observed in the modified cantilevers. (c) The modified cantilevers also exhibit significantly smaller Brownian fluctuations (~7–11 pN) while stretching the unfolded polyprotein, while the unmodified cantilever showed resonant oscillations during this time (d) that are apparent in its power spectrum density (PSD). Thus, FIB-modified cantilevers can show microsecond response times without underdamped oscillations. (Adapted with permission from Edwards, D.T., et al., *Nano Lett.*, 15, 7091–7098, 2015.)

have refolded by that specific time interval (Figure 3.8a). (Interestingly, sometimes the refolding of a protein can occur with such strong propensities that the refolding events will register as a "deflection" of the cantilever during the reapproach [Bornschlögl and Rief 2006; Lee et al. 2006; Kim et al. 2010; Lee et al. 2010; Zheng et al. 2014]). Hence, refolding experiments can be done in a

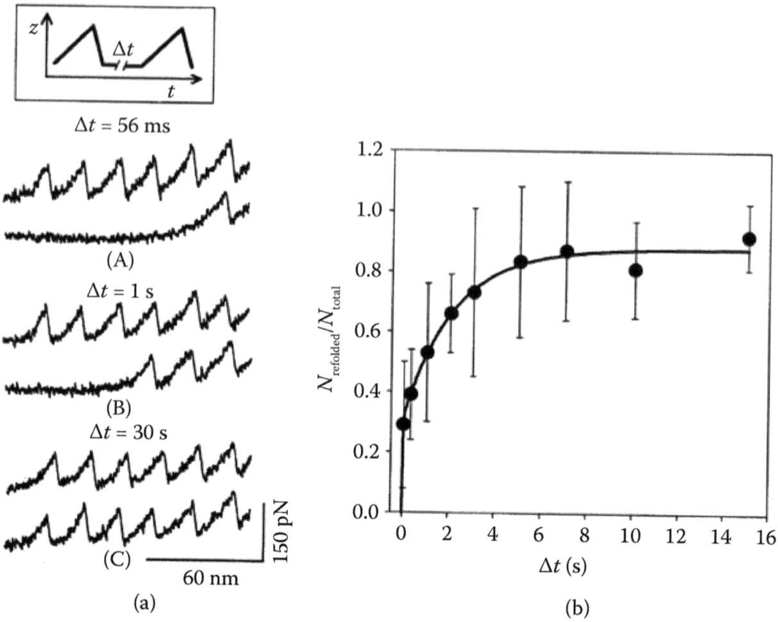

Figure 3.8 Determination of the folding rates of a polyprotein of tenascin, a protein of the extracellular matrix. (a) After unfolding the polyprotein initially, the cantilever is brought back to the surface so the applied forces are lowered for a prescribed period of time (Δt), then the tip is retracted again. During this time, individual tenascin domains may have refolded, and their subsequent unfolding events will then be observed in the next *F–X* curve. If a tenascin domain has not yet refolded, then in the *F–X* curve it will behave simply as an unstructured polypeptide chain. (b) From the fraction of domains observed for different Δt's, the rate of protein folding can be determined (in addition to the rates of protein unfolding). (Adapted by permission from Macmillan Publishers Ltd., *Nature,* Oberhauser, A.F., et al., 393, 181–185, 1998. Copyright © [1998].)

similar manner to unfolding experiments. Often these experiments are done with a polyprotein containing several repeats of the same domain of interest so that "multiple" refolding experiments can be conducted from one *F–X* curve.

After conducting this experiment many times, on many molecules, and at different time intervals, the intrinsic folding rate of protein k_f can be determined (Figure 3.8b). Together with the experimentally determined intrinsic unfolding rate k_u of a protein (described previously), the folding free energy profile of a protein ΔG_f or of each of its domains can be experimentally determined with SMFS.

3.3.3 AFM SMFS Experiments to Probe Receptor–Ligand Binding

A significant application of SMFS is also the direct measurement of receptor–ligand or protein–protein (e.g., antigen–antibody) interactions (Kienberger et al. 2006; Zhang et al. 2009; Noy 2011; Willet et al. 2012) (Figure 3.9). This can be

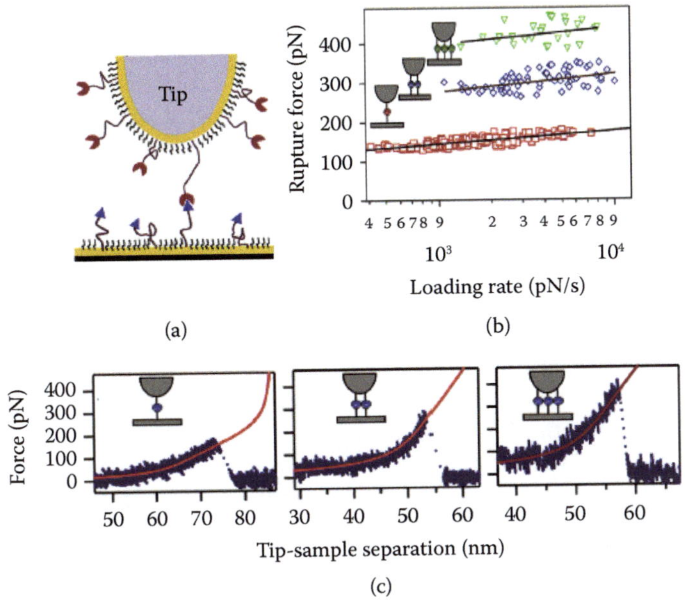

Figure 3.9 (a) A receptor–ligand SMFS experiment. The AFM tip is coated with a ligand while the surface is coated with the antigen via long linkers. The density of the molecules of interest is kept low, in order to maximize the probability that only a single ligand-antigen pair will interact during a given approach of the cantilever toward the surface. (b) Example of the rupture force versus loading rates, from which the energetic parameters of the interactions can be derived. The occasional measurements of multiple bonds rather than individual bonds can be easily differentiated. (c) Examples of *F–X* curves for different numbers of receptor–ligand binding events. (Reprinted from *Curr. Opin. Chem. Biol.*, 15, Noy, A., 710–718, 2011. Copyright © (2011), with permission from Elsevier.)

achieved by depositing one molecule of interest on the surface, while chemically tethering its binding partner directly to the AFM tip (Figure 3.9a). For these experiments, control of the binding pose or relative orientations of the interacting biomolecules while they are being manipulated is of critical importance. This control can be achieved through careful selection of the attachment chemistries between the biomolecules and either the surface or AFM tip (such as those described above), or the addition of a long "linker" molecule to attach the molecules to the tip or surface (Figure 3.9a). These linkers can provide additional flexibility for the ligand and antigen to attain binding poses similar to what they may attain in solution. Once the molecules of interest have been attached with their respective chemistries, these experiments can be performed in a similar manner as are the protein unfolding experiments with repeated approach/retraction of the cantilever toward the surface. Often the tip must be incubated near the surface for a relatively long period of time (a few seconds) in order for a ligand to find and bind to an antigen with a reasonably high probability.

The peaks in the *F–X* curves should now be considered as the “rupture” force (force at which the ligands and antigens dissociate). After the completion of many of these experiments, the distribution of observed rupture forces can be assembled into a histogram. The presence of multimodal distributions in the binding histogram may suggest that more than one ligand–antigen pair was being manipulated for different *F–X* curves (Figure 3.9b and c). Furthermore, it is critical that control experiments be performed in the absence of one or both members of the binding pair in order to identify the distribution of apparent rupture forces that result from nonspecific interactions with the surface rather than the interaction of interest.

The forces of unbinding at different pulling rates can be related to the binding energies, in a similar manner as to what was previously described (Figure 3.9b); however, in cases where binding and mechanical unbinding may be reversible over the course of the experiment (such as when pulling is relatively slow), care must be taken in the interpretation of the rupture force distributions, such as described by Friddle, Noy, and De Yoreo (Friddle et al. 2012). Interestingly, this technique can be combined with AFM imaging modalities to map the location of receptors in synthetic membranes and even on living cells and simultaneously record their interactions with ligands that have been attached to the AFM tip (Muller et al. 2009). Recently, this technique was exploited to measure the kinetic and thermodynamic properties of ligand binding for protease-activated receptor-1 (PAR1) molecules (Alsteens et al. 2015) and experimentally reconstruct the ligand-binding free energy landscape using the Friddle–Noy–De Yoreo model.

3.4 Highlights of Recent Applications of AFM Force Spectroscopy in Answering Biological Questions

Having discussed some of the technical and practical details of the AFM SMFS, in this section, we highlight some recent applications of SMFS to answer poignant biological questions that would be nearly impossible to address in the absence of this versatile technique.

3.4.1 Mechanically Activated ATP Binding of a Strain-Sensing Kinase Domain of Titin

As mentioned earlier in the chapter, the mechanical properties of the Fn and Ig domains from the muscle protein titin have been well characterized using force spectroscopy. However, the titin protein also contains kinase domains whose response to force was not well understood. This titin kinase was thought to be a biological strain sensor—that is, it was thought to transfer phosphate from intracellular ATP molecules to other biological molecules in response to a mechanical strain, which could have the effect of regulating the transcription of muscle genes and the expression or degradation of proteins in muscle cells in response to externally applied forces. Although it was found that partial denaturation of the titin kinase could stimulate its phosphorylation activity for obvious reasons biochemical experiments could not directly observe the proposed force-dependent activation of the kinase domain in order to test this hypothesis.

Puchner et al. (2008) performed a series of AFM experiments that allowed them to directly probe the propensities for the titin kinase to bind ATP molecules in response to applied forces. Using AFM SMFS, they mechanically denatured purified segments of the titin protein containing the titin kinase and its naturally surrounding Ig and Fn domains (Figure 3.10a). After multiple repetitions of the experiment, they could identify a mechanical fingerprint of five distinct events assigned to the unfolding of the titin kinase domain (Figure 3.10b). When ATP was added to the solution (Figure 3.10c), a number of the *F–X* curves exhibited an additional peak that was not observed in experiments that lacked ATP (peak 2*). The probability that this peak was observed increased with increasing ATP concentration (Figure 3.10d) and decreased significantly if a lysine residue in the binding site of the kinase, which was predicted to interact with the ATP, was mutated to an alanine (Figure 3.10e); these results would strongly suggest that the new peak indicated a structural transition that occurred in response to the binding of ATP by the kinase, which could be resolved during its mechanical unfolding.

Furthermore, in combination with molecular dynamics and biochemical experiments, Puchner et al. found that the probability of ATP binding was inversely proportional to the loading rate ν. Together with computational studies, these results suggested that mechanical forces could perturb the structure of the kinase domain to open an autoinhibited ATP-binding site to allow for it to bind to ATP (Figure 3.10a). Thus, using AFM SMFS in combination with computational and biochemical experiments, the mechanically activated functionality of the titin kinase could be experimentally verified and its mechanisms characterized.

3.4.2 Force-Clamp Technique to Differentiate the Chemical Mechanisms of Disulfide Reduction by Divergent Thioredoxin Proteins

Thioredoxins are enzymes that catalyze the reduction of disulfide (S–S) bonds, which can form between two cysteine (Cys) residues, and are involved in a multitude of cellular processes. These enzymes are evolutionarily widespread, being present in bacteria as well as in higher organisms through plants and mammals, but possess a conserved Cys-X-X-Cys motif in their active site. Biochemical data suggested that the reduction of disulfide bonds by thioredoxins proceeded via an S_N2 mechanism (i.e., a bimolecular nucleophilic substitution reaction). However, why different thioredoxins catalyzed this reaction at different rates, and how these differences depended on the specific amino acid sequence and structure of the different thioredoxins, remained unclear (Perez-Jimenez et al. 2009). The Fernandez group, using SMFS using an AFM as a force clamp, was able to monitor the states of individual disulfide bonds in the presence of thioredoxin—first to characterize the chemical mechanisms of bacterial (*Escherichia coli*) thioredoxin, and then a number of evolutionary divergent thioredoxins. Using this technique, they could identify and differentiate a number of catalytic mechanisms by their modulation of different forces applied to the disulfide bonds. Together with computational modeling, the origin of the diversity of reaction rates and mechanisms by thioredoxins could be directly investigated (four bacterial-origin thioredoxins, and four eukaryotic-origin thioredoxins).

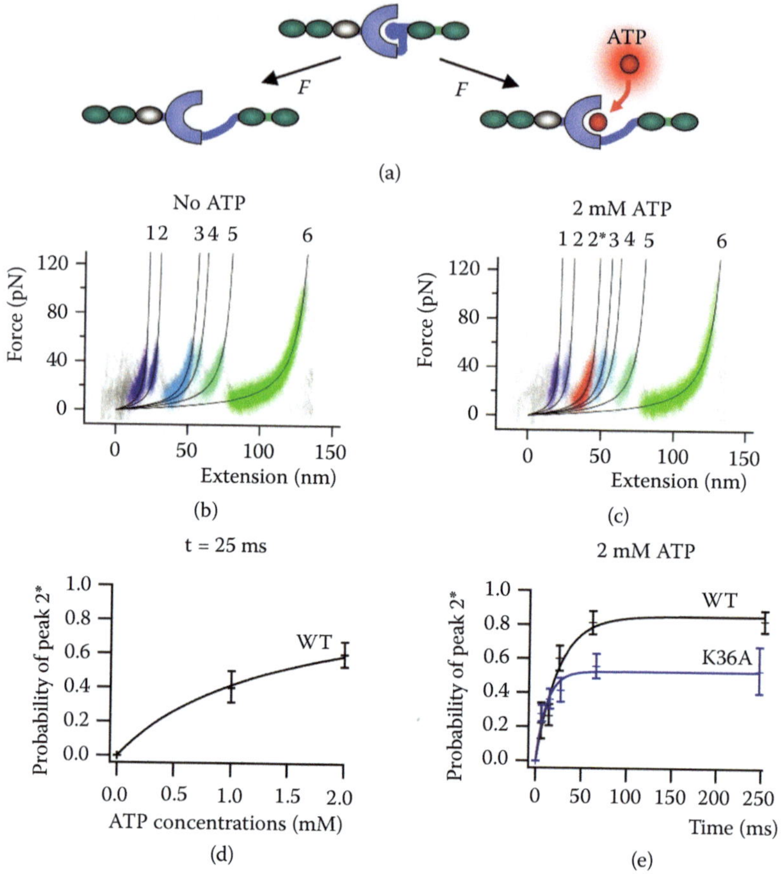

Figure 3.10 (a) Titin kinase (flanked by titin Ig and Fn domains) is autoinhibited in the absence of force. When subjected to applied mechanical forces, it was suspected that a cryptic ATP binding domain/active site is revealed, commencing its phosphorylation activity. (b) In the absence of ATP, the *F*–*X* curve of the titin kinase exhibits five unfolding events, while (c) in the presence of ATP, a new unfolding event (2*) is observed as unfolding between Events 2 and 3. (d) The probability of observing peak 2* increased with increasing ATP concentration, and (e) mutation of a lysine to an alanine (K36A) in the predicted active site of the kinase reduced the probability that this event was observed. Slowing the pulling speed (and therefore increasing the amount of time during which the ATP binding site is exposed and when event 2* would be recorded) increases the probability that this event is observed as well. (Reprinted from Puchner, E.M., et al., *Proc. Natl. Acad. Sci. USA*, 105, 13385–13390, 2008. With permission. Copyright © [2008] National Academy of Sciences, U.S.A.)

In order to be certain that the states of individual disulfide bonds were being monitored, internal cysteine residues were introduced into the I27 domains of an I27 polyprotein (Figure 3.11a). Initially, large forces were applied to an individual polyprotein to denature and extend all of the domains, with the exception of a protected segment of the I27s that was mechanically insulated by the internal S–S bond. As a result, a polyprotein with intact internal S–S bonds and one whose S–S bond had been reduced could be differentiated by their two different contour lengths. After this rapid unfolding of the I27s, the applied force was jumped to one of several forces (Figure 3.11b) in order to measure the effect of these applied forces on the catalytic reduction of the S–S bonds (Figure 3.11c and d).

For pea chloroplastic thioredoxin (Figure 3.11c) and other bacterial-origin thioredoxins, as well as for popular thioredoxin *Trxh1* (Figure 3.11d) and other eukaryotic-origin thioredoxins, low applied forces decreased the rate of S–S reduction. This behavior is consistent with a Michaelis–Menten type S_N2 reaction whose rate is limited by the rate of proper binding of the S-S substrate into the active site of the thioredoxin—that is, a specific geometry between the active site cysteines of the thioredoxin and the S–S bond must be achieved for catalytic reduction to occur efficiently, and the applied forces decreased the probability that this geometry could be achieved. At higher forces, the behavior of the bacterial-origin and the eukaryotic-origin thioredoxins diverged: after a minimum in catalytic rate around 200 pN, the bacterial-origin thioredoxins exhibited a subsequent increase in catalytic rate with increasing forces applied across the S–S bond of its substrate, while the eukaryotic-origin thioredoxins reduced the S–S bonds with a nonzero, force-independent rate. After fitting the reaction kinetics of the bacterial-origin thioredoxins to a force-dependent model of disulfide reduction (similar as to what was described above), the experimentally derived location of the S–S transition state $x^{\ddagger}$ was found to be very close to what would be expected for the location of the transition state for a simple S_N2 reaction, such as what would occur for the reduction of S–S bonds in the presence of free cysteine molecules rather than within a structurally well-defined active site. The force-independent catalytic reduction of S–S bonds exhibited by eukaryotic-origin thioredoxins was found to result from a single-electron transfer mechanism via electron tunneling, which likely also occurs with bacterial-origin thioredoxins but is obscured by the much faster S_N2 mechanism. Using computational modeling of the structures of the different thioredoxins, it was found that this propensity to perform the S_N2 mechanism in bacterial-origin thioredoxins was highly correlated with a shallower active site, which allowed for the S_N2 reaction to occur. Hence, this ability to apply forces across individual S–S bonds and monitor their behavior allowed for diverse chemical mechanisms by thioredoxin and their origins to be unraveled.

3.4.3 AFM Force Spectroscopy Techniques to Study the Folding and Refolding of Large, Multidomain Proteins: The N-Terminal Domain of Luciferase Controls Misfolding Avoidance

Single-domain proteins make up less than half the proteomes in all kingdoms of life, yet they are the best-characterized proteins because of the relative ease

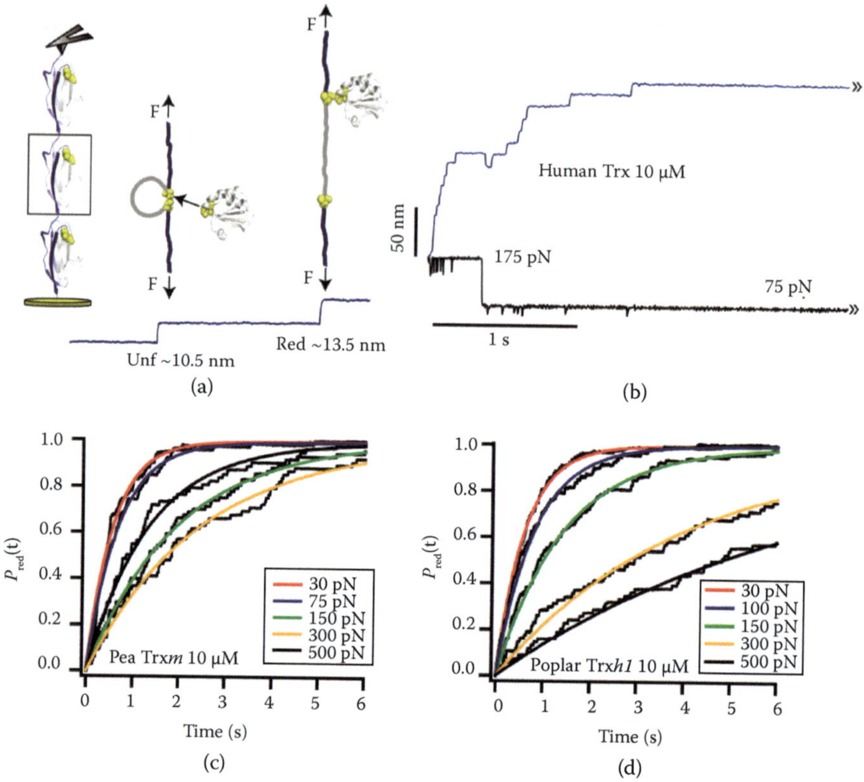

Figure 3.11 (a) A force clamp experiment to probe the catalytic mechanism of thioredoxins. A polyprotein of I27 domains containing a buried disulfide (S–S) bond is subjected to high forces to denature the I27 domains and expose the S–S bonds. The force is then dropped, and the reduction of the S–S bond by thioredoxins in solution is monitored by the change in the length of the polyprotein at that force. (b) Example force spectrogram of the polyprotein extended in the presence of human thioredoxin (Trx). The blue trace is the recorded length of the polyprotein over time, and the black trace is the applied force over time. (c) While the rate of reduction by thioredoxin for bacterial origin, like pea mitochondrial thioredoxin (Trxm), decreases with increasing applied forces and then begins to increase again, (d) for the eukaryotic-origin thioredoxins like poplar thioredoxin, the rate decreases and then remains approximately constant at higher forces (see Perez-Jimenez et al. [2009] for more examples of evolutionarily divergent thioredoxins). The rates of reduction by thioredoxin are estimated from the single molecule data by finding $1/\tau$, where τ is the exponential time constant derived from the fit (black line) to the experimentally observed probability that a reduction event occurred before time t ($P_{red}(t)$). (Adapted by permission from Macmillan Publishers Ltd., *Nat. Struct. Mol. Biol.*, Perez-Jimenez, R., et al., 16, 890–896, 2009. Copyright © [2009].)

with which they can be studied using traditional biochemical methods (Han et al. 2007). Multidomain proteins, however, are not nearly as well characterized, and they can exhibit behaviors that deviate significantly from single-domain proteins—for example, multidomain proteins have been shown to undergo kinetic partitioning (Peng and Li 2008), cotranslational folding (Frydman et al. 1999), or require chaperone proteins for folding (Hartl et al. 2011). AFM force spectroscopy is able to circumvent many of the experimental obstacles in studying multidomain proteins, such as protein aggregation, and allows for precise characterization of the unfolding and refolding of multidomain proteins.

A recent study by Scholl et al. (2014) looked at the unfolding and refolding of luciferase—one of the largest proteins yet studied by AFM SMFS. Luciferase is a 550 residue protein that is responsible for the enzymatic reaction that allows fireflies to produce light. It had been shown previously to fold cotranslationally (Frydman et al. 1999), and under conditions used to study luciferase with biochemical techniques, luciferase cannot refold and will aggregate without the aid of chaperones (Herbst et al. 1998). After mechanically unfolding the protein, Scholl et al. used simulations and repeated SMFS experiments with truncated mutants to directly assign three peaks in the luciferase force versus extension profile (Figure 3.12a) to the forced unfolding of the C-terminal domain, the middle domain, and the N-terminal domain, respectively, from first to last unfolding events (Figure 3.12a inset). Scholl et al. then tested the ability of luciferase to refold after fully unfolding luciferase (pulling the protein to full extension). After a full extension then retraction cycle, the original peaks could not be recovered on subsequent unfolding pulls (Figure 3.12b)—instead large, inconsistent peaks that behaved as would be expected for misfolded structures were observed—unless the proteins were supplemented with a buffer containing chaperones (Figure 3.12c). While this result was consistent with the previous biochemical experiments, Scholl et al. then tested the ability of luciferase domains to refold after only partially unfolding the entire protein (pulling the protein to partial extension), which corresponded to unfolding only the C-terminal domain or only the C-terminal and middle domain of luciferase (Figure 3.12d). These experiments uncovered something new—they showed that luciferase could refold with similar propensity as with chaperones if only partially unfolded, indicating that the folded N-terminal domain provides a benefit for luciferase refolding. SMFS studies of folding and refolding of multidomain proteins like this provide a means to untangle the interactions between protein domains and understand their influence on protein structure and stability.

3.4.4 Uncovering Hidden Interactions Governing Amyloid β-Protein Dimerization Using AFM Force Spectroscopy

Alzheimer's disease is typically caused by plaques formed by the aggregation of amyloid β-proteins (Aβ) (Roychaudhuri et al. 2009). Because Aβ proteins are created during the cleavage of amyloid precursor proteins, Aβ proteins come in various, similar forms of approximately 40 amino acids. The most abundant form is Aβ40 (which is 40 amino acids in size) but the most toxic form is Aβ42, which only has two additional C-terminal residues compared to Aβ40. The reasons for the increased toxicity of Aβ42 resulting from this small difference in size are still not understood.

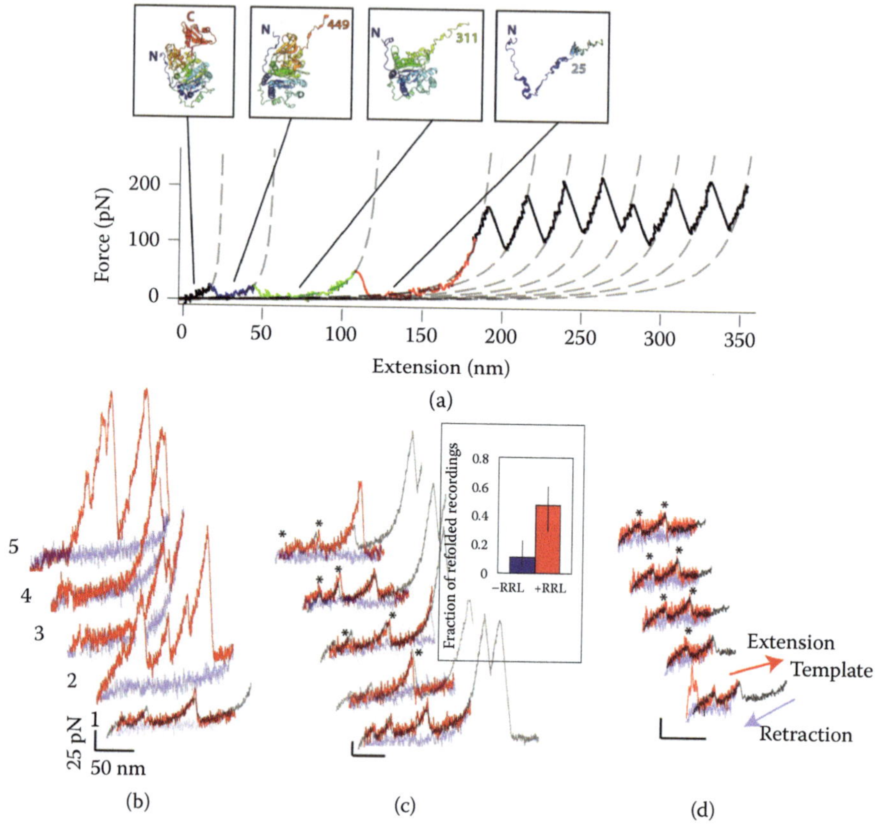

Figure 3.12 (a) The force–extension of a polyprotein composed of three N-terminal I27 domains followed by luciferase and four additional I27 is shown. The curve first shows three peaks corresponding to the unfolding of three domains of luciferase, followed by the seven peaks from the unfolding of each of the seven I27 domains. The I27 domains are mechanically stronger, which is why they always appear after the peaks resulting from the luciferase. The boxes at the top show a molecular dynamics simulation of the unfolding of luciferase providing evidence that the unfolding of luciferase occurs from the C-terminal domain to the N-terminal domain. See Scholl et al. (2014) for additional experimental evidence for this observation. (b–d) Unfolding and refolding are performed on luciferase by extending the molecule (red) and then retracting (blue) and then applying a subsequent extension–retraction pulse (traces labeled 2 through 5). (b) On its own, the luciferase protein is seen not to be able to refold after being fully unfolded. (c) With the addition of chaperone proteins (from rabbit reticulocyte lysate), luciferase proteins are observed to refold ~50% of the time. (d) When there is no chaperone, the rest of the luciferase protein can refold ~70% of the time if its N-terminal domain remained folded during these experiments. (Adapted from Scholl, Z.N., Yang, W.T., and Marszalek, P.E., *J. Biol. Chem.*, 289, 28607–28618, 2014.)

The Lyubchenko group used AFM SMFS to examine the differences between Aβ40 and Aβ42 in their binding energy and the nature of their interactions.

Lv et al. (2013) attached Aβ peptides directly to an AFM tip and then to a surface in order to measure the interaction between the two proteins (Figure 3.13 schematic). As expected, Aβ42 dimers were found to have stronger interpeptide

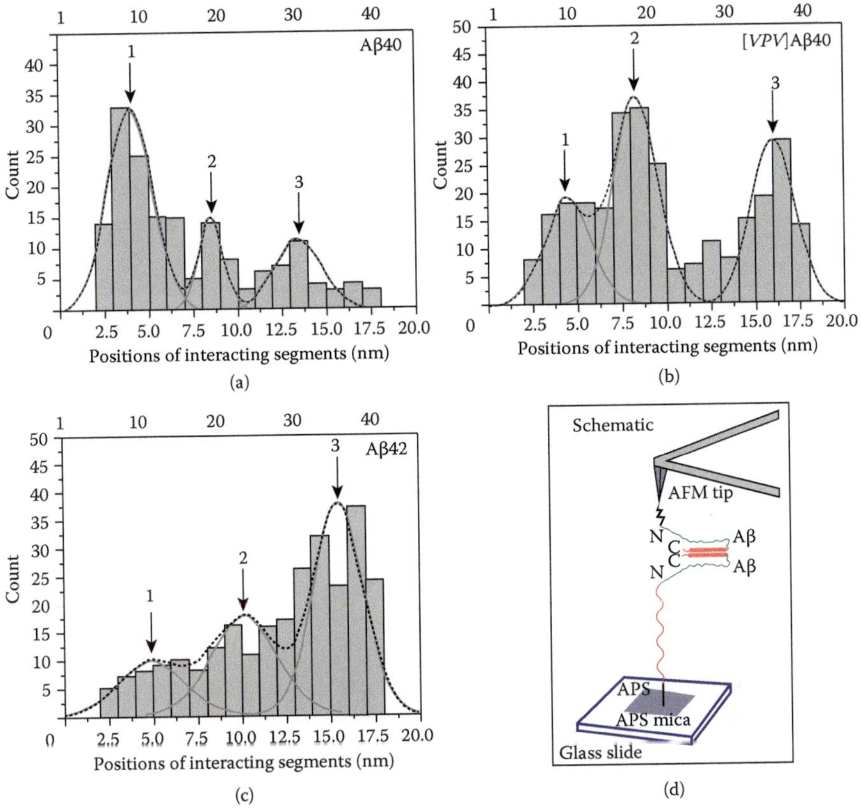

Figure 3.13 Measuring the extent of interaction between dimers of Aβ amyloid proteins (experimental schematic bottom right). By fitting the *F*–*X* curve of the dimeric polymer being stretched, the lengths of the C-terminal segments of the Aβ peptides that are forming the dimeric interaction are determined. The upper *X*-axis shows the residue number of the first interacting amino acids in the dimers (from their N-termini). Aβ40 and Aβ42 exhibit three distinct interactive conformations (Peaks 1, 2, and 3). (a) The Aβ40 dimers most often observed have large regions of interactions stretching from their C-termini to 1–10 residues from their N-termini (Peak 1). (b) An N-terminal mutation (VPV) makes this interaction less likely, while also increasing the strength of the interaction between dimers (see Lv et al. [2013] for more details)—in effect, (c) making its interactions more like Aβ42, (d) schematic representation of the interaction between the two proteins. (Adapted from Lv, Z., Roychaudhuri, R., Condron, M.M., Teplow, D.B., and Lyubchenko, Y.L., *Sci. Rep.*, 3, 2013. With permission.)

interactions than Aβ40 dimers. Lv et al. then studied the extent of interaction between the monomeric peptides of the dimers. After subtracting the length of the linkers, the contour lengths of the N-terminal segments of the Aβ peptides which were not interacting as part of Aβ-Aβ dimer were experimentally determined (Figure 3.13a through c). This allowed Lv et al. to determine the length of the Aβ peptides that were in fact part of a dimeric structure. They found that both Aβ40 dimers and Aβ42 dimers adopted three dominant conformations with different lengths of their regions of interactions. For dimers of Aβ40, the most commonly observed interaction was one that contained a relatively long interaction stretching from its C-terminus to the 1–10 residues from its N-terminus (peak 1 in Figure 3.13a), while the most common dimers of Aβ42 possessed interactions localized mostly to its C-terminus (Peak 3 in Figure 3.13c). This finding demonstrated that the two additional amino acids of Aβ42 can dramatically alter the nature of its interaction with other Aβ peptides.

The introduction of a VPV mutation (Gly33Val-Val36Pro-Gly38Val) into the Aβ40 peptide ([VPV]Aβ40), which is known to increase the toxicity of Aβ40 and make it more "Aβ42-like," was also found to increase the rupture forces and lifetime of the interpeptide interactions between Aβ40 monomers. At the same time, the distribution of [VPV]Aβ40 dimers observed was shifted to those with shorter C-terminal interactions. Thus, even though the main difference between the Aβ42 and Aβ40 amyloid proteins is the C-terminal residues, based on the results of the SMFS experiments it appears that the N-terminal region may play an important role in interpeptide interactions, which may make it a potential target to prevent Aβ aggregation.

References

Alegre-Cebollada, J., Perez-Jimenez, R., Kosuri, P. & Fernandez, J. M. 2010. Single-molecule force spectroscopy approach to enzyme catalysis. *J Biol Chem* 285, 18961–18966.

Alsteens, D., Pfreundschuh, M., Zhang, C., Spoerri, P. M., Coughlin, S. R., Kobilka, B. K., Muller, D. J. 2015. Imaging G protein-coupled receptors while quantifying their ligand-binding free-energy landscape. *Nat Meth* 12, 845–851.

Baumann, F., Bauer, M. S., Milles, L. F., Alexandrovich, A., Gaub, H. E. & Pippig, D. A. 2015. Monovalent Strep-Tactin for strong and site-specific tethering in nanospectroscopy. *Nat Nano* 11, 89–94.

Bertz, M. & Rief, M. 2009. Ligand binding mechanics of maltose binding protein. *J Mol Biol*, 393, 1097–1105.

Bornschlögl, T. & Rief, M. 2006. Single molecule unzipping of coiled coils: Sequence resolved stability profiles. *Phys Rev Lett* 96, 118102.

Bouchiat, C., Wang, M. D., Allemand, J., Strick, T., Block, S.M. & Croquette, V. 1999. Estimating the persistence length of a worm-like chain molecule from force-extension measurements. *Biophys J* 76, 409–413.

Bustamante, C., Chemla, Y.R., Forde, N. R. & Izhaky, D. 2004. Mechanical processes in biochemistry. *Ann Rev Biochem* 73, 705–748.

Bustamante, C., Marko, J. F., Siggia, E. D. & Smith, S. 1994. Entropic elasticity of lambda-phage DNA. *Science* 265, 1599–1600.

Butt, H. J., Cappella, B. & Kappl, M. 2005. Force measurements with the atomic force microscope: Technique, interpretation and applications. *Surf Sci Rep* 59, 1–152.

Cao, Y. & Li H. 2007. Polyprotein of GB1 is an ideal artificial elastomeric protein. *Nat Mater* 6, 109–114.

Carrion-Vazquez, M., Oberhauser, A. F., Fowler, S. B., Marszalek, P. E., Broedel, S. E., Clarke, J. & Fernandez, J. M. 1999. Mechanical and chemical unfolding of a single proteIn: A comparison. *Proc Natl Acad Sci U S A* 96, 3694–3699.

Clausen-Schaumann, H., Rief, M. & Gaub, H. E. 1999. Sequence dependent mechanics of single DNA molecules. *Biophys J*, 76, A151.

Dietz, H. & Rief, M. 2006. Protein structure by mechanical triangulation. *Proc Natl Acad Sci*, 103, 1244–1247.

Dougan, L. & Fernandez, J. M. 2007. Tandem repeating modular proteins avoid aggregation in single molecule force spectroscopy experiments. *J Phys Chem A*, 111, 12402–12408.

Dudko, O. K., Hummer, G. & Szabo, A. 2006. Intrinsic rates and activation free energies from single-molecule pulling experiments. *Phys Rev Lett*, 96, 108101.

Dufrêne, Y. & Hinterdorfer, P. 2008. Recent progress in AFM molecular recognition studies. *Pflügers Archiv Eur J Physiol*, 456, 237–245.

Edwards, D. T., Faulk, J. K., Sanders, A. W., Bull, M. S., Walder, R., Leblanc, M.-A., Sousa, M. C. & Perkins, T. T. 2015. Optimizing 1-μs-resolution single-molecule force spectroscopy on a commercial atomic force microscope. *Nano Lett*, 15, 7091–7098.

Evans, E., Kinoshita, K., Simon, S. & Leung, A. 2010. Long-lived, high-strength states of ICAM-1 bonds to β 2 integrin, I: Lifetimes of bonds to recombinant α L β 2 under force. *Biophys J*, 98, 1458–1466.

Evans, E. & Ritchie, K. 1997. Dynamic strength of molecular adhesion bonds. *Biophys J*, 72, 1541–1555.

Fernandez, J. M. & Li, H. 2004. Force-clamp spectroscopy monitors the folding trajectory of a single protein. *Science*, 303, 1674–1678.

Fisher, T. E., Marszalek, P. E. & Fernandez, J. M. 2000. Stretching single molecules into novel conformations using the atomic force microscope. *Nat Struct Mol Biol*, 7, 719–724.

Fisher, T. E., Oberhauser, A. F., Carrion-Vazquez, M., Marszalek, P. E. & Fernandez, J. M. 1999. The study of protein mechanics with the atomic force microscope. *Trends Biochem Sci*, 24, 379–384.

Florin, E., Moy, V. & Gaub, H. 1994. Adhesion forces between individual ligand-receptor pairs. *Science*, 264, 415–417.

Florin, E. L., Rief, M., Lehmann, H., Ludwig, M., Dornmair, C., Moy, V. T. & Gaub, H. E. 1995. Sensing specific molecular-interactions with the atomic-force microscope. *Biosens Bioelectron*, 10, 895–901.

Flory, P. J. 1953. *Principles of polymer chemistry.* Cornell University Press, Ithaca, NY.

Friddle, R. W., Noy, A. & De Yoreo, J. J. 2012. Interpreting the widespread nonlinear force spectra of intermolecular bonds. *Proc Natl Acad Sci*, 109, 13573–13578.

Frydman, J., Erdjument-Bromage, H., Tempst, P. & Hartl, F. U. 1999. Co-translational domain folding as the structural basis for the rapid de novo folding of firefly luciferase. *Nat Struct Biol*, 6, 697–705.

Han, J.-H., Batey, S., Nickson, A. A., Teichmann, S. A. & Clarke, J. 2007. The folding and evolution of multidomain proteins. *Nat Rev Mol Cell Biol*, 8, 319–330.

Hanley, W., MCcarty, O., Jadhav, S., Tseng, Y., Wirtz, D. & Konstantopoulos, K. 2003. Single molecule characterization of P-selectin/ligand binding. *J Biol Chem*, 278, 10556–10561.

Hartl, F. U., Bracher, A. & Hayer-Hartl, M. 2011. Molecular chaperones in protein folding and proteostasis. *Nature*, 475, 324–332.

He, C., Hu, C., Hu, X., Hu, X., Xiao, A., Perkins, T. T. & LI, H. 2015. Direct observation of the reversible two-state unfolding and refolding of an α/β protein by single-molecule atomic force microscopy. *Angewandte Chemie Int Edn*, 54, 9921–9925.

Herbst, R., Gast, K. & Seckler, R. 1998. Folding of firefly (*Photinus pyralis*) luciferase: Aggregation and reactivation of unfolding intermediates. *Biochemistry*, 37, 6586–6597.

Hinterdorfer, P. & Dufrene, Y. F. 2006. Detection and localization of single molecular recognition events using atomic force microscopy. *Nat Meth*, 3, 347–355.

Hummer, G. & Szabo, A. 2001. Free energy reconstruction from nonequilibrium single-molecule pulling experiments. *Proc Natl Acad Sci U S A*, 98, 3658–3661.

Isralewitz, B., Gao, M., Lu, H. & Schulten, K. 2000. Forced unfolding or mutant titin immunoglobulin domains with steered molecular dynamics. *Biophys J*, 78, 28a.

Junker, J. P., Ziegler, F. & Rief, M. 2009. Ligand-dependent equilibrium fluctuations of single calmodulin molecules. *Science*, 323, 633–637.

Kellermayer, M. S. Z., Smith, S. B., Bustamante, C. & Granzier, H. L. 1998. Complete unfolding of the titin molecule under external force. *J Struct Biol*, 122, 197–205.

Kessler, M., Gottschalk, K. E., Janovjak, H., Muller, D. J. & Gaub, H. E. 2006. Bacteriorhodopsin folds into the membrane against an external force. *J Mol Biol*, 357, 644–654.

Kienberger, F., Ebner, A., Gruber, H. J. & Hinterdorfer, P. 2006. Molecular recognition imaging and force spectroscopy of single biomolecules. *Acc Chem Res*, 39, 29–36.

Kim, B.-H., Palermo, N. Y., Lovas, S. N., Zaikova, T., Keana, J. F. W. & Lyubchenko, Y. L. 2011. Single-molecule atomic force microscopy force spectroscopy study of Aβ-40 interactions. *Biochemistry*, 50, 5154–5162.

Kim, M., Abdi, K., Lee, G., Rabbi, M., Lee, W., Yang, M., Schofield, C. J., Bennett, V. & Marszalek, P. E. 2010. Fast and forceful refolding of stretched alpha helical solenoid proteins. *Biophys J*, 98, 3086–3092.

Lee, G., Abdi, K., Jiang, Y., Michaely, P., Bennett, V. & Marszalek, P. E. 2006. Nanospring behaviour of ankyrin repeats. *Nature*, 440, 246–249.

Lee, G. U., Chrisey, L. A. & Colton, R. J. 1994. Direct measurement of the forces between complementary strands of DNA. *Science*, 266, 771–773.

Lee, W., Zeng, X., Yang, W. & Marszalek, P. E. 2012. Mechanics of polysaccharides. *In:* Anne-Sophie Duwez, N. W. (ed.), *Molecular manipulation with atomic force microscopy*. Boca Raton, FL: CRC Press.

Lee, W., Zeng, X., Zhou, H.-X., Bennett, V., Yang, W. & Marszalek, P. E. 2010. Full reconstruction of a vectorial protein folding pathway by atomic force microscopy and molecular dynamics simulations. *J Biol Chem*, 285, 38167–38172.

Li, H. B., Linke, W. A., Oberhauser, A. F., Carrion-Vazquez, M., Kerkviliet, J. G., Lu, H., Marszalek, P. E. & Fernandez, J. M. 2002. Reverse engineering of the giant muscle protein titin. *Nature*, 418, 998–1002.

Li, H. B., Rief, M., Oesterhelt, F. & Gaub, H. E. 1998. Single-molecule force spectroscopy on Xanthan by AFM. *Adv Mater*, 10, 316–319.

Li, Q., Scholl, Z. N. & Marszalek, P. E. 2014. Capturing the mechanical unfolding pathway of a large protein with coiled-coil probes. *Angewandte Chemie-Int Edn*, 53, 13429–13433.

Love, J. C., Estroff, L. A., Kriebel, J. K., Nuzzo, R. G. & Whitesides, G. M. 2005. Self-assembled monolayers of thiolates on metals as a form of nanotechnology. *Chem Rev*, 105, 1103–1169.

Ludwig, M., Moy, V. T., Rief, M., Florin, E. L. & Gaub, H. E. 1994. Characterization of the Adhesion force between Avidin-functionalized Afm tips and biotinylated agarose beads. *Microscopy Microanal Microstruct*, 5, 321–328.

Lv, Z., Roychaudhuri, R., Condron, M. M., Teplow, D. B. & Lyubchenko, Y. L. 2013. Mechanism of amyloid [bgr]-protein dimerization determined using single-molecule AFM force spectroscopy. *Sci Rep*, 3, 2880.

Marko, J. F. & Siggia, E. D. 1995. Stretching DNA. *Macromolecules*, 28, 8759–8770.

Marszalek, P. E. & Dufrene, Y. F. 2012. Stretching single polysaccharides and proteins using atomic force microscopy. *Chem Soc Rev*, 41, 3523–3534.

Marszalek, P. E., Li, H. B. & Fernandez, J. M. 2001. Fingerprinting polysaccharides with single-molecule atomic force microscopy. *Nat Biotechnol*, 19, 258–262.

Marszalek, P. E., Oberhauser, A. F., Pang, Y. P. & Fernandez, J. M. 1998. Polysaccharide elasticity governed by chair-boat transitions of the glucopyranose ring. *Nature*, 396, 661–664.

Marszalek, P. E., Pang, Y. P., Li, H. B., El Yazal, J., Oberhauser, A. F. & Fernandez, J. M. 1999. Atomic levers control pyranose ring conformations. *Proc Natl Acad Sci U S A*, 96, 7894–7898.

Merkel, R., Nassoy, P., Leung, A., Ritchie, K. & Evans, E. 1999. Energy landscapes of receptor-ligand bonds explored with dynamic force spectroscopy. *Nature*, 397, 50–53.

Muller, D. J. 2008. AFM: A nanotool in membrane biology. *Biochemistry*, 47, 7986–7998.

Muller, D. J., Helenius, J., Alsteens, D. & Dufrene, Y. F. 2009. Force probing surfaces of living cells to molecular resolution. *Nat Chem Biol*, 5, 383–390.

Muller, D. J., Heymann, J. B., Oesterhelt, F., Moller, C., Gaub, H., Buldt, G. & Engel, A. 2000. Atomic force microscopy of native purple membrane. *Biochim Biophys Acta-Bioenerg*, 1460, 27–38.

Noy, A. 2011. Force spectroscopy 101: How to design, perform, and analyze an AFM-based single molecule force spectroscopy experiment. *Curr Opin Chem Biol*, 15, 710–718.

Oberhauser, A. F. & Carrion-Vazquez, M. 2008. Mechanical biochemistry of proteins one molecule at a time. *J Biol Chem*, 283, 6617–6621.

Oberhauser, A. F., Hansma, P. K., Carrion-Vazquez, M. & Fernandez, J. M. 2001. Stepwise unfolding of titin under force-clamp atomic force microscopy. *Proc Natl Acad Sci U S A*, 98, 468–472.

Oberhauser, A. F., Marszalek, P. E., Erickson, H. P. & Fernandez, J. M. 1998. The molecular elasticity of the extracellular matrix protein tenascin. *Nature*, 393, 181–185.

Oesterhelt, F., Oesterhelt, D., Pfeiffer, M., Engel, A., Gaub, H. E. & Muller, D. J. 2000. Unfolding pathways of individual bacteriorhodopsins. *Science*, 288, 143–146.

Peng, Q. & Li, H. 2008. Atomic force microscopy reveals parallel mechanical unfolding pathways of T4 lysozyme: Evidence for a kinetic partitioning mechanism. *Proc Natl Acad Sci*, 105, 1885–1890.

Perez-Jimenez, R., Garcia-Manyes, S., Ainavarapu, S. R. K. & Fernandez, J. M. 2006. Mechanical unfolding pathways of the enhanced yellow fluorescent protein revealed by single molecule force spectroscopy. *J Biol Chem*, 281, 40010–40014.

Perez-Jimenez, R., Li, J., Kosuri, P., Sanchez-Romero, I., Wiita, A. P., Rodriguez-Larrea, D., Chueca, A., et al. 2009. Diversity of chemical mechanisms in thioredoxin catalysis revealed by single-molecule force spectroscopy. *Nat Struct Mol Biol*, 16, 890–896.

Phillips, J. C., Braun, R., Wang, W., Gumbart, J., Tajkhorshid, E., Villa, E., Chipot, C., Skeel, R. D., Kalé, L. & Schulten, K. 2005. Scalable molecular dynamics with NAMD. *J Comput Chem*, 26, 1781–1802.

Popa, I., Kosuri, P., Alegre-Cebollada, J., Garcia-Manyes, S. & Fernandez, J. M. 2013. Force dependency of biochemical reactions measured by single-molecule force-clamp spectroscopy. *Nat Prot*, 8, 1261–1276.

Pronk, S., Pall, S., Schulz, R., Larsson, P., Bjelkmar, P., Apostolov, R., Shirts, M. R., et al. 2013. GROMACS 4.5: A high-throughput and highly parallel open source molecular simulation toolkit. *Bioinformatics*, 29, 845–854.

Puchner, E. M., Alexandrovich, A., Kho, A. L., Hensen, U., Schafer, L. V., Brandmeier, B., Grater, F., Grubmuller, H., Gaub, H. E. & Gautel, M. 2008. Mechanoenzymatics of titin kinase. *Proc Natl Acad Sci U S A*, 105, 13385–13390.

Puchner, E. M. & Gaub, H. E. 2012. Single-molecule mechanoenzymatics. *Ann Rev Biophys*, 41, 497–518.

Rabbi, M. & Marszalek, P. 2008. Probing polysaccharide and protein mechanics by atomic force microscopy. In: Selvin, P. R. & Ha, T. (eds.) *Single-molecule techniques: A laboratory manual*. Cold Spring Harbor: Cold Spring Harbor Laboratory Press.

Rico, F., Gonzalez, L., Casuso, I., Puig-Vidal, M. & Scheuring, S. 2013. High-speed force spectroscopy unfolds titin at the velocity of molecular dynamics simulations. *Science*, 342, 741–743.

Rief, M., Gautel, M., Oesterhelt, F., Fernandez, J. M. & Gaub, H. E. 1997a. Reversible unfolding of individual titin immunoglobulin domains by AFM. *Science*, 276, 1109–1112.

Rief, M., Oesterhelt, F., Heymann, B. & Gaub, H. E. 1997b. Single molecule force spectroscopy on polysaccharides by atomic force microscopy. *Science*, 275, 1295–1297.

Roychaudhuri, R., Yang, M., Hoshi, M. M. & Teplow, D. B. 2009. Amyloid beta-protein assembly and Alzheimer disease. *J Biol Chem*, 284, 4749–4753.

Schlierf, M., Berkemeier, F. & Rief, M. 2007. Direct observation of active protein folding using lock-in force spectroscopy. *Biophys J*, 93, 3989–3998.

Scholl, Z. N., Li, Q. & Marszalek, P. E. 2013a. Single molecule mechanical manipulation for studying biological properties of proteins,DNA, and sugars. *WIREs Nanomed Nanobiotechnol*, 2, 211–229.

Scholl, Z. N., Rabbi, M., Lee, D., Manson, L., S-Gracz, H. & Marszalek, P. E. 2013b. Origin of overstretching transitions in single-stranded nucleic acids. *Phys Rev Lett*, 111, 188302.

Scholl, Z. N., Yang, W. & Marszalek, P. E. 2014. Chaperones rescue luciferase folding by separating its domains. *J Biol Chem*, 289, 28607–28618.

Scholl, Z. N., Yang, W. T. & Marszalek, P. E. 2015. Direct observation of multimer stabilization in the mechanical unfolding pathway of a protein undergoing oligomerization. *ACS Nano*, 9, 1189–1197.

Schuler, B. & Hofmann, H. 2013. Single-molecule spectroscopy of protein folding dynamics—Expanding scope and timescales. *Curr Opin Struct Biol*, 23, 36–47.

Schwaiger, I., Kardinal, A., Schleicher, M., Noegel, A. A. & Rief, M. 2004. A mechanical unfolding intermediate in an actin-crosslinking protein. *Nat Struct Mol Biol*, 11, 81–85.

Shlyakhtenko, L. S., Gall, A. A., Filonov, A., Cerovac, Z., Lushnikov, A. & Lyubchenko, Y. L. 2003. Silatrane-based surface chemistry for immobilization of DNA, protein–DNA complexes and other biological materials. *Ultramicroscopy*, 97, 279–287.

Sotomayor, M. & Schulten, K. 2007. Single-molecule experiments in vitro and in silico. *Science*, 316, 1144–1148.

Stigler, J. & Rief, M. 2012. Calcium-dependent folding of single calmodulin molecules. *Proc Natl Acad of Sci*, 109, 17814–17819.

Stigler, J., Ziegler, F., Gieseke, A., Gebhardt, J. C. M. & Rief, M. 2011. The complex folding network of single calmodulin molecules. *Science*, 334, 512–516.

Sullan, R. M. A., Churnside, A. B., Nguyen, D. M., Bull, M. S. & Perkins, T. T. 2013. Atomic force microscopy with sub-pico Newton force stability for biological applications. *Methods*, 60, 131–141.

Taniguchi, Y. & Kawakami, M. 2010. Application of HaloTag protein to covalent immobilization of recombinant proteins for single molecule force spectroscopy. *Langmuir*, 26, 10433–10436.

Ulman, A. 1996. Formation and structure of self-assembled monolayers. *Chem Rev*, 96, 1533–1554.

Willet, N., Lamprecht, C., Rankl, C., Rangl, M., Creasey, R., Ebner, A., Voelcker, N. H. & Hinterdorfer, P. 2012. Molecular recognition force spectroscopy. In: Anne-Sophie Duwez, N. W. (ed.), *Molecular manipulation with atomic force microscopy*. Boca Raton, FL: CRC Press.

Williams, P. M., Fowler, S. B., Best, R. B., Toca-Herrera, J. L., Scott, K. A., Steward, A. & Clarke, J. 2003. Hidden complexity in the mechanical properties of titin. *Nature*, 422, 446–449.

Wojcikiewicz, E. P., Abdulreda, M. H., Zhang, X. & Moy, V. T. 2006. Force spectroscopy of LFA-1 and its ligands, ICAM-1 and ICAM-2. *Biomacromolecules*, 7, 3188–3195.

Zhang, X., Rico, F., Xu, A. J. & Moy, V. T. 2009. Atomic force microscopy of protein–protein interactions handbook of single-molecule biophysics. In: Hinterdorfer, P. & Oijen, A. (eds.). Handbook of Single-Molecule Biophysics, pp. 555–570 Springer US, New York.

Zhang, Y. & Lyubchenko, Y. L. 2014. The structure of misfolded amyloidogenic dimers: Computational analysis of force spectroscopy data. *Biophys J*, 107, 2903–2910.

Zheng, P., Wang, Y. & Li, H. 2014. Reversible unfolding–refolding of rubredoxin: A single-molecule force spectroscopy study. *Angewandte Chem Int Edn*, 53, 14060–14063.

Zhuang, X. W. & Rief, M. 2003. Single-molecule folding. *Curr Opin Struct Biol*, 13, 88–97.

4 Magnetic Tweezers

Piero R. Bianco
University at Buffalo
Buffalo, New York

Yuri L. Lyubchenko and Zhiqiang Sun
University of Nebraska Medical Center
Omaha, Nebraska

Contents

4.1 Introduction

Mechanical force is an important component of life implicit in molecular, cellular, and organismal function. Recent advanced biophysics research has demonstrated the ability to measure these forces at the single molecule level (reviewed in Tanase et al. 2007; Lionnet et al. 2012). In order to probe the interactions between molecules and thereby provide insight into biomechanical mechanism, we need to have sensitive tools. A wide variety of force-probing techniques, including atomic force microscopy (AFM) and optical and magnetic tweezers, have been developed. The first two methods are presented in this book. Briefly, in AFM, the cantilever is used to apply force to a molecule by either pulling or pushing. In contrast, the tweezers-based approaches use either tightly focused laser beams or magnetic fields to "tweeze or trap" objects. The history of single molecule magnetic tweezers methodology is presented in box 4.1 Displacement from the initial point in the trap as a function of opposing or assisting force is then used to provide distance measurements and additional insight, such as the spacing of mechanical steps. This chapter focuses on magnetic tweezers, where a "magnetic trap" is used to exert force on superparamagnetic beads that are coupled to single molecules.

Small magnetic probes have been used in biological studies for years to measure forces controlling cell growth and differentiation as well as forces required for remodeling of the extracellular matrix. The aim of this chapter is to present the principles underlying magnetic tweezers for probing single molecules, along with advantages and limitations of the method.

4.2 Principle of Magnetic Tweezers

Magnetic tweezers belong to the family of single molecule manipulation instruments capable of probing molecules over a broad range of forces from 0.01 to 10^4 pN (Neuman and Nagy 2008). As is clear from the method's name, magnetic tweezers utilize a magnetic field to manipulate individual molecules (Strick et al. 1996).

Box 4.1 History of Single Molecule Magnetic Tweezers Methodology

The first single molecule probing experiments in magnetic field were described in Smith et al. (1992). In these experiments, lambda DNA was immobilized at one end to the glass surface, and magnetic bead was attached to another end of DNA. The magnetic field generated by a magnet mounted above the sample was applied to stretch DNA molecules, and these experiments were used to characterize mechanical properties of DNA. These pioneering studies prompted Strick et al. (1996) to develop a magnetic trap instrument by which they were able to manipulate DNA attached to the magnetic bead. Importantly, they were able to coil the DNA molecule in a controllable fashion by rotating the magnet and also to stretch the DNA by moving the magnet in the vertical direction.

4.2.1 Instrument Configuration and Design

A schematic of the magnetic tweezers instrument capable of manipulating the DNA molecule is shown in Figure 4.1. The setup of magnetic tweezers is presented in Box 4.2. One end of DNA is covalently immobilized on the lower, coverslip surface of a flow cell mounted on the top of inverted microscope. A paramagnetic bead (0.5–4.5 μm in diameter) is attached to the other end of the DNA. Small magnets placed above the flow cell generate a magnetic field by which the paramagnetic bead is trapped, so the bead can be moved in 3D and rotate. As a result, the DNA molecule can be stretched and/or twisted. The observation of the bead motion is performed via the inverted optical microscope on which the magnetic tweezers apparatus is integrated. Stretching forces as large as ~100 pN and as small as 0.01 pN can be measured by the instrument. Rotating magnets generate torque as strong as 10^4–10^5 nm/rad (De Vlaminck and Dekker 2012). Torque in this range is sufficient to rotate DNA in the direction of the double

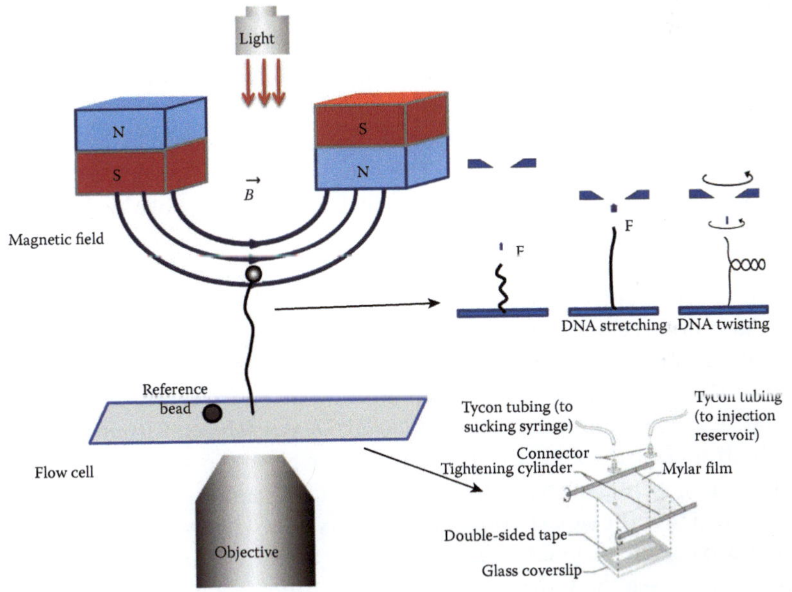

Figure 4.1 Magnetic tweezers. A bead is immobilized to a functionalized surface of the flow cell via a DNA tether. The flow cell is mounted on top of an inverted microscope that allows one to follow the bead position. A permanent magnet mounted above the cell generates the magnetic field that creates the driving force to the bead. The inset illustrates the major operation of the instrument. At small distance between the bead and the magnet, the magnetic field is strong enough that the bead can move vertically (stretching of the tether; middle image) or the bead can rotate, twisting the tether (right picture). The motion of the bead is observed through the objective and a CCD camera in real time. (Reprinted with permission from Lionnet, T., et al., *Cold Spring Harb. Protoc.*, 34, 2012.)

Box 4.2 Specifics to the Magnetic Tweezers Setup

According to Figure 4.1, a regular magnetic tweezers instrument is mounted on the top of an inverted microscope with the light coming from the top. Typically, a parallel light beam is generated by a regular LED light source and filtered by a field lens.

To generate a strong magnetic field, the NdFeB magnets are normally used in the experiments. As in Figure 4.1, two pieces of magnets are mounted close to the sample, and the magnets are coupled with a small motor that controls the magnets' position, moving them up and down and rotating them as required by the experiment. Magnet movement is translated to microbead allowing the experimenter to stretch the DNA molecule by pulling the bead up and twisting the DNA, applying rotation to the magnet. It is important for these manipulations to center the magnet pair relative to the optical axis; therefore, the bead is centered relative to the optical beam of the microscope to allow the experimenter to accurately measure the bead position during the rotation of the magnets. The gap between the magnets and the sample should be kept as small as possible, as the magnetic field decays rapidly with the distance in vertical direction. Typically, this distance is ~2 mm.

The sample is mounted inside the flow cell that should be thin to keep the magnets close to the sample. Typically, the flow chamber is formed by a thin Mylar film and glass coverslip, which are glued together by the double-side stick tape. The buffer can be exchanged through the two holes on the top film. With this setup, the magnet can be as close as 200 μm to the bead, allowing a strong magnetic field to be applied to the microbead. The microbead is not free but is tethered to the inner surface by a tether—long DNA molecules in the majority of cases of DNA. Generally, the bead is coated with streptavidin, which can specifically bind to the biotin-labeled DNA end; at the same time, the other end of DNA was labeled with digoxigenin, which specifically and strongly binds to the coverslip surface coated with anti-digoxigenin antibodies.

The size of the superparamagnetic beads varied between 0.5 and 4.5 μm. For a constant magnetic field, increasing the radius (R) of the magnetic bead increases the force substantially; the amplification factor is proportional to R^3. When the application requires large force (>50 pN), relatively large beads are needed—Dynabead M-280 (2.8 μm diameter) and M-450 (4.5 μm diameter) being the most popular choices. However, the large bead also had some disadvantages. The Brownian forces acting on the beads increase when the bead diameter increases (linear to $\sqrt{R}$), and the signal-to-noise ratio for some particular trapping experiments decreases (linear to $1/\sqrt{R}$). So, if the large external force is not required, the MyOne bead (1 μm diameter) will be a good choice.

The bead position and movement are controlled by the optical microscope and captured by a CCD camera attached to one of the ports of the microscope. For a short DNA (3 μm contour length) attached to a 2.8 μm bead and the applied force around 10 pN, the frequency of fluctuation is in the order of 10 Hz. In such cases, a low-end CCD camera is capable of recording all the fluctuation accurately. However, for a high-resolution experiment requiring a subnanometer resolution, a 100 Hz acquisition camera is needed, DNA as small as 1 μm is used, and the applied force can be larger than 3 pN.

helix or in the opposite direction to generate positive or negative DNA supercoiling, respectively (Strick et al. 1996). The magnetic bead plays a dual role. First, it is a force transducer to the system, such as the DNA molecule in Figure 4.1. Second, the bead position defines the x, y, and z coordinates by which parameters such as the extension of the tether is determined. In the schematic shown, the tether is a single molecule of DNA. Details of the measurements are explained in the subsections that follow.

4.2.2 Magnetic Trap

Here, we provide the basics for magnetic trap operation. The force F generated by the external magnetic field is proportional to the gradient (∇) of the magnetic field $\vec{B}$ and is, therefore, directed toward the area with the strongest field (Lipfert et al. 2009). The force applied to the magnetic probe is defined from the following equation:

$$\vec{F} = \frac{1}{2} \cdot \vec{\nabla}\left(\overrightarrow{mo} \cdot \vec{B}\right) \tag{4.1}$$

where $\vec{mo}$ is the induced magnetic moment of the bead in the magnetic field $\vec{B}$. As a result, magnetic tweezers display different properties depending on the configuration of the field. As the magnetic field is almost constant in the millimeter scale, and both bead size and movement are in the micron to submicron range, the magnetic field for the range of bead movement can be considered uniform , so $\vec{B}$ in Equation 4.1 is constant (Vilfan et al. 2009). This is an important factor that considerably simplifies the data analysis and theoretical basis for the method.

Manipulation by magnets changes the bead position, so determination of the bead position is the most critical factor in a magnetic tweezers experiment. In the majority of instruments (Figure 4.1), the bead is illuminated by a parallel light beam, so the scattered and incident beams interfere, producing a diffraction pattern (Gosse and Croquette 2002). In these calibration experiments, the objective focus is moved in a vertical direction, and the diffraction patterns corresponding to selected distances are recorded. Typical images are shown in Figure 4.2a through d. This diffraction pattern imaged by a high-resolution CCD camera is

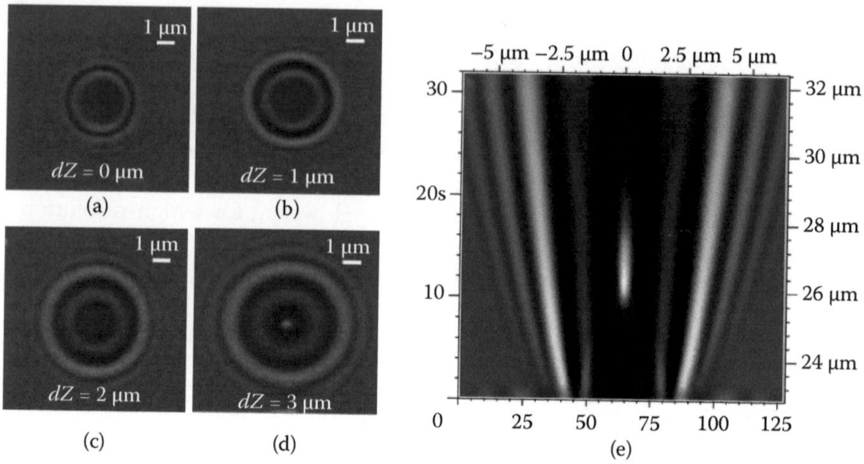

Figure 4.2 Generation of a calibration profile in the z direction for a bead in magnetic tweezers. A 10 μm × 10 μm image of the diffraction pattern of a reference bead with a diameter of ~3 μm at different positions of the objective focal plane (OFP). First, the OFP is positioned close to the bead (a). Next, the OFP is moved away from the bead, and the images of the diffraction rings are recorded at successive positions of the OFP (b–d). If the bead is illuminated with a parallel light source, the diffraction pattern of the bead will always be identical when the objective is focused at the same relative distance from the bead. By recording the size and intensity of the diffraction rings at different positions of the OFP, a calibration profile can be made (e), which can be used to measure vertical relative displacements of the bead with an accuracy of ~10 nm, limited by the resolution of the piezo that controls the position of OFP. (Reprinted with permission from Vilfan, I.D., et al., *Handbook of Single Molecule Biophysics*, pp. 371, Springer-Verlag, New York, 2009.)

then captured by a frame grabber board in a computer. These data are assembled to generate a calibration graph (Figure 4.2e). Here, the bead position can be determined with an accuracy of ~10 nm. If, in addition to the tethered bead, a bead attached directly to the surface is used (i.e., reference bead), the diffraction method allows one to directly measure the length of the tether. The use of the reference bead also helps in eliminating the drift of the sample stage relative to the light beam.

4.2.3 Force Measurements with Magnetic Tweezers

The major feature of magnetic tweezers is that the force measurements are done directly without any calibration of the sensor. This property is the inherent feature of the instrument in which force is measured from analysis of the Brownian motion of the tethered bead. As shown in Figure 4.3, thermal motion displaces the bead from its initial position, but restoring force exerted by the tether

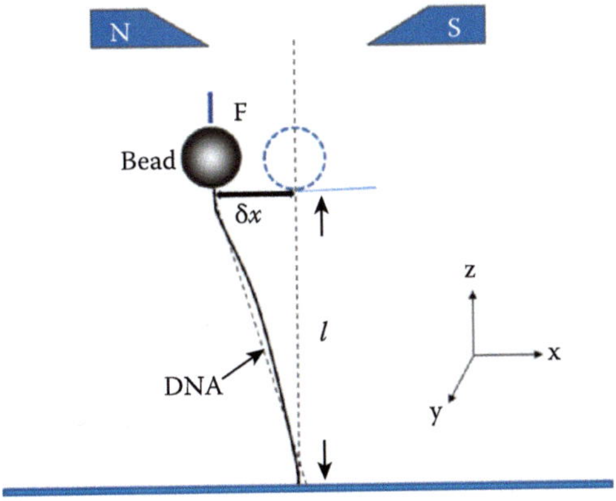

Figure 4.3 Brownian motion of the tethered bead in solution.

(e.g., DNA molecule) works against this displacement. Analysis of this motion (Lionnet et al. 2012) leads to the following expression for the dependence of force F on the range of the bead fluctuation along the y-axis δy attached to the tether with length l:

$$F = \frac{kTl}{<\delta y^2>} \tag{4.2}$$

Graphically, the Brownian motion of the bead is illustrated in Figure 4.4a, in which red and green data points correspond to experiments at low and high forces, respectively. Figure 4.4b demonstrates the time trajectory of the bead motion along the x-axis with the fluctuation spectrum to the right. Such data obtained at different forces are assembled as graphs in Figure 4.4c (blue curve). The force-dependent extension of the tether l is measured by the diffraction method as described above. In fact, Equation 4.2 is limited to relatively low forces. The use of Fourier analysis extends the analysis at higher forces (Vilfan et al. 2009), and it is presented in Box 4.3.

The spatial resolution for magnetic tweezers is defined by two factors—thermal noise and the instrument-dependent resolution in tracking the bead. Details for the thermal noise limitation are given by De Vlaminck and Dekker (2012). All factors are taken into consideration, and according to this analysis, a subnanometer resolution is possible with magnetic tweezers, although the applied force should be above 3 pN. The instrument-dependent resolution limits are determined by the CCD and bead-illumination characteristics. Note a recent paper in which subnanometer resolution has been achieved (Dulin et al. 2015). To do this, a bright laser was used to track the bead with 6 kHz frequency. As a result, 0.5 nm steps were recorded for the bead tethered by a double-stranded DNA, and the DNA hairpin dynamics on the millisecond timescale were detected.

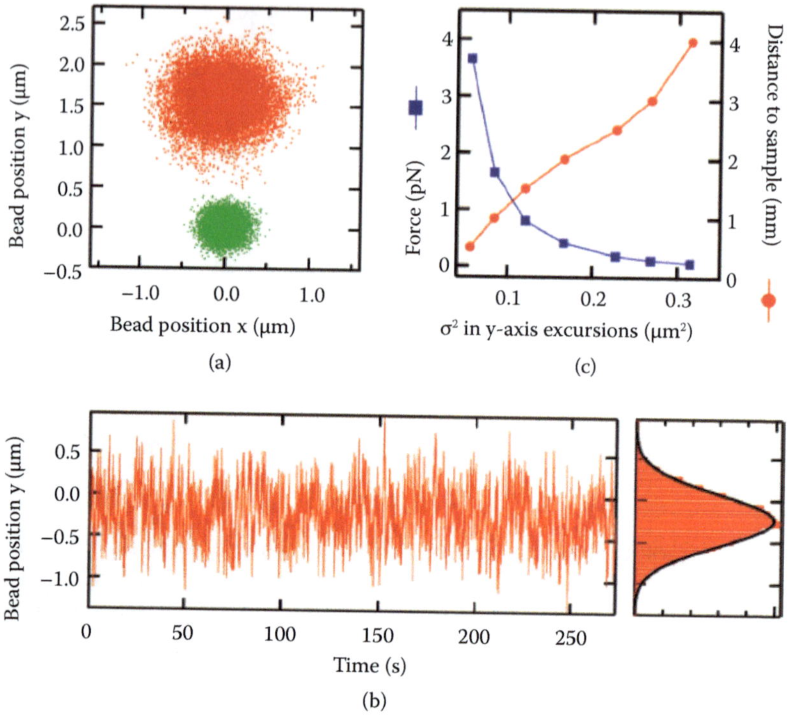

Figure 4.4 Measuring the applied force using the fluctuations in bead excursions in real space. (a) The excursions of the bead center in the (x, y) plane for low force (*red*) and higher force (*green*). The data points for the low force have been offset for better representation. (b) The bead excursion in the x dimension in time (the low-force measurement from panel [a] is shown), with a histogram on the right, including a Gaussian fit to the histogram. (c) The computed variance of the bead excursions (not obtained from a fit to the Gaussian) for a variety of magnet positions (*red solid circles*), as well as the calculated corresponding forces (*blue solid squares*). (Reprinted with permission from Vilfan, I. D., et al., *Handbook of Single Molecule Biophysics*, pp. 371, Springer-Verlag, New York, 2009.)

4.2.4 Torque Measurements with Magnetic Tweezers

Magnetic beads are not perfectly isotropic in shape; therefore, anisotropy of the bead magnetization is the factor that is responsible for the bead rotation (Lipfert et al. 2009). If $\vec{m}$ is the magnetic moment of the bead not parallel to the direction of magnetic field $\vec{B}$, the torque $\vec{\Gamma}$ is defined by the following equation:

$$\vec{\Gamma} = \vec{m} X \vec{B} \tag{4.3}$$

Torque measurement with magnetic tweezers is one of the attractive features of this method. One of the examples enabling direct generation of positive and

Box 4.3 Analysis of the Brownian Motion of the Bead at High Force

Typically, the applied force can be calculated by Equation 4.2 in real space, but the $<\delta x^2>$ is limited by the systematic error, which is much more obvious at high force. The systematic error limitation can come from the sample drift or the finite integration time of the CCD camera, and the effects of these errors arise when the applied force increases. To filter out the noise from the Brownian motion, the computation of $<\delta x^2>$ can be performed in Fourier space, which allows for a more accurate determination.

In the Fourier space, $<\delta x^2>$ can be calculated by integrating the power spectrum of the bead motion:

$$<\delta x^2>=\frac{1}{2\pi}\int P(\omega)d\omega \tag{4.4}$$

where P(ω) is the power spectrum of the bead motion (Vilfan et al. 2009). Combining the power spectrum in a regular experiment and introducing the Fourier transform, by defining $\omega_c = k_x/6\pi\eta R$,

$$P(\omega)=\frac{2k_B T 6\pi\eta R}{\left(1+\left(\frac{\omega}{\omega_c}\right)^2\right)\pi k_x^2} \tag{4.5}$$

where R is the radius of the bead and $k_x = F/l$, η is the viscosity of the buffer.

Plugging Equation 4.4 into Equation 4.5 leads to the following equation for $<\delta x^2>$:

$$<\delta x^2>=\int_0^\infty \frac{24\pi k_B TR}{\pi k_x^2}\frac{1}{1+(\omega/\omega_c)^2}d\omega \tag{4.6}$$

This equation can be used for the calculation of forces using Equation 4.2. Details for analysis with the use of this approach can be found in Vilfan et al. (2009).

negative supercoiling in a DNA duplex is described in Section 4.2.5. Details for setting up the experiments with bead rotation and data analyses along with video demonstrations are provided in Lipfert et al. (2009).

The bead rotation studies require precise and gentle control of torque applied to the system (Figure 4.5a). Note that due to the large difference in lateral sizes of the bead and the DNA molecule, the torque generated by rotating the bead is several orders of magnitude larger than the torque applied to the DNA molecule (De Vlaminck and Dekker 2012). Therefore, control of DNA rotation is needed in these experiments.

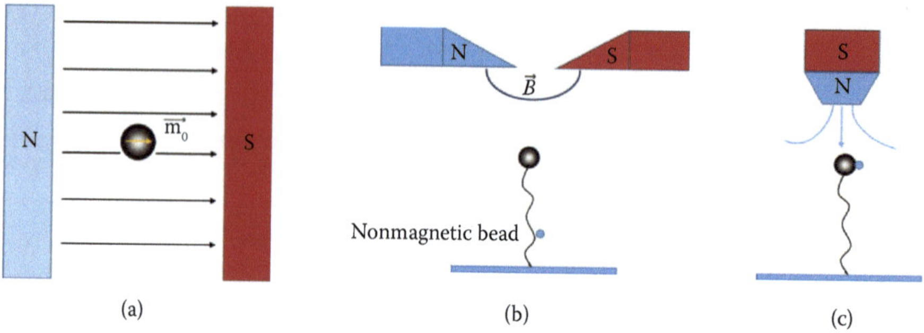

Figure 4.5 Measurements of torque with magnetic tweezers. (a) Top view of the conventional implementation of magnetic tweezers. The magnetic field is shown by horizontal lines and the bead is oriented along the magnetic field lines. The side view of the experimental setup is shown in (b), where the magnetic bead is tethered by DNA to the glass substrate. (c) Side views of configurations that rely on circularly symmetric fields generated by a single magnet.

In one approach, a nonmagnetic bead is attached to DNA that allows one to directly follow the rotation of DNA in the images captured by the CCD camera (Figure 4.5b). Such a torque-and-twist-sensitive magnetic tweezers was originally proposed by Gore et al. (2006) and was applied to measure the twist-stretch coupling in single DNA molecules. Another approach for more sensitive control of torque is to utilize a magnet with a cylindrical configuration field (Figure 4.5c). This field configuration is generated by the cylindrical magnets, which produce a magnetic field that is circularly symmetric about the central axis of the magnet (Celedon et al. 2009). In this configuration, a magnetic bead is aligned to the center of magnet, and only a soft torque applied to the bead is produced. A similar magnet configuration was used in Lipfert et al. (2010) to precisely track the angle of the bead with high precision ($\delta\theta \approx 0.1°$). This was achieved by attaching a small bead marker as shown in Figure 4.5b. This high-precision level of torque measurement was critical in measuring the torsional stiffness of RecA filaments, which is nearly twofold larger than that of dsDNA. Experimental details of these magnetic tweezers experiments with a focus on twist and torque measurements can be found in Lipfert et al. (2014).

4.2.5 DNA Supercoiling Probed by Measurements of Torque

A key feature of magnetic tweezers is their ability to apply torque to a tethered molecule by simply rotating the magnet shown in Figure 4.1. This capability was demonstrated for the first time in Strick et al. (1996). Here, magnetic tweezers were used to twist DNA molecules to generate molecules with positive and negative superhelical turns. More on DNA supercoiling and topology is presented under Box 4.4.

These types of DNA manipulations were first demonstrated in 1996 (Strick et al. 1996). Schematically the experiments are depicted in Figure 4.6a. First, a stretching force F is applied to the bead to straighten the DNA as shown in

Box 4.4 DNA Supercoiling and Topology

A DNA molecule can be characterized by its linking number Lk_0, which is the ratio between the number of DNA base pairs (N) and the number of base pairs in the helical turn γ, so $Lk_0 = N/\gamma$. The change of number to value Lk, for example, by twisting in the direction opposite to the right-hand DNA helix, leads to the linking number deficit $\Delta Lk = Lk - Lk_0$. Due to the stiffness of the DNA duplex, this linking number deficit leads to duplex overwinding, termed *DNA supercoiling*; therefore, the ratio $\sigma = \Delta Lk/Lk_0$ is termed *DNA supercoiling* (e.g., Sinden et al. 1980; Bliska and Cozzarelli 1987). This value is negative in the example considered above because the torque was applied to the direction opposite of the right-handed DNA duplex geometry, so $Lk - Lk_0$ is negative. Rotation of the duplex in the opposite direction overwinds DNA, so $Lk - Lk_0 > 0$; such a DNA molecule is termed a *positively supercoiled DNA*. Note that DNA *in vivo* is negatively supercoiled.

cartoon (i). Second, the magnet is rotated and, consequently, the bead rotates. Such rotation leads to the formation of DNA supercoils as depicted in scheme (ii). As a result of this conformational transition, the bead moves in a downward direction toward the coverslip surface, and this DNA compaction is detected by the bead position measured by the tweezers. Clockwise rotation is used for generating positive supercoils, and counterclockwise rotation produces negatively supercoiled DNA.

Experimental results depicted as the dependence of DNA extension as a function of turns for a typical supercoiling experiment are shown in Figure 4.6b (diamond symbols). This is a bell-shaped curve with the maximum extension at zero rotation and is symmetric relative to the rotation direction. However, it is valid for stretching forces smaller than 0.5 pN. The picture changes if a larger stretching force is applied. The results for 1 pN stretching force are shown in Figure 4.6b (squares). The bead does not change the position if negative supercoiling is induced (counterclockwise rotation). Only rotation as large as ~0.03 positive supercoiling density results in the DNA molecule becoming supercoiled. No supercoiling is observed when a force as large as 8 pN is applied to the DNA duplex as evidenced by a constant bead position (circles in Figure 4.6b).

This phenomenon was investigated in a subsequent study by the same group (Strick et al. 1998), and the explanation is that stretching force can induce structural transition in DNA double helix. Under stretching forces in the range between 0.5 and 3 pN, the negative turn can result in untwisting and melting of DNA double helix that is accompanied by a barely detectable change of the DNA extension. On the contrary, overwound DNA formed under positive rotation of the bead becomes positively supercoiled, and the number of these turns increases upon bead rotation, resulting in a decrease in DNA extension. At large stretching forces (>6.5 pN), the DNA duplex is fully destroyed and the bead rotation is

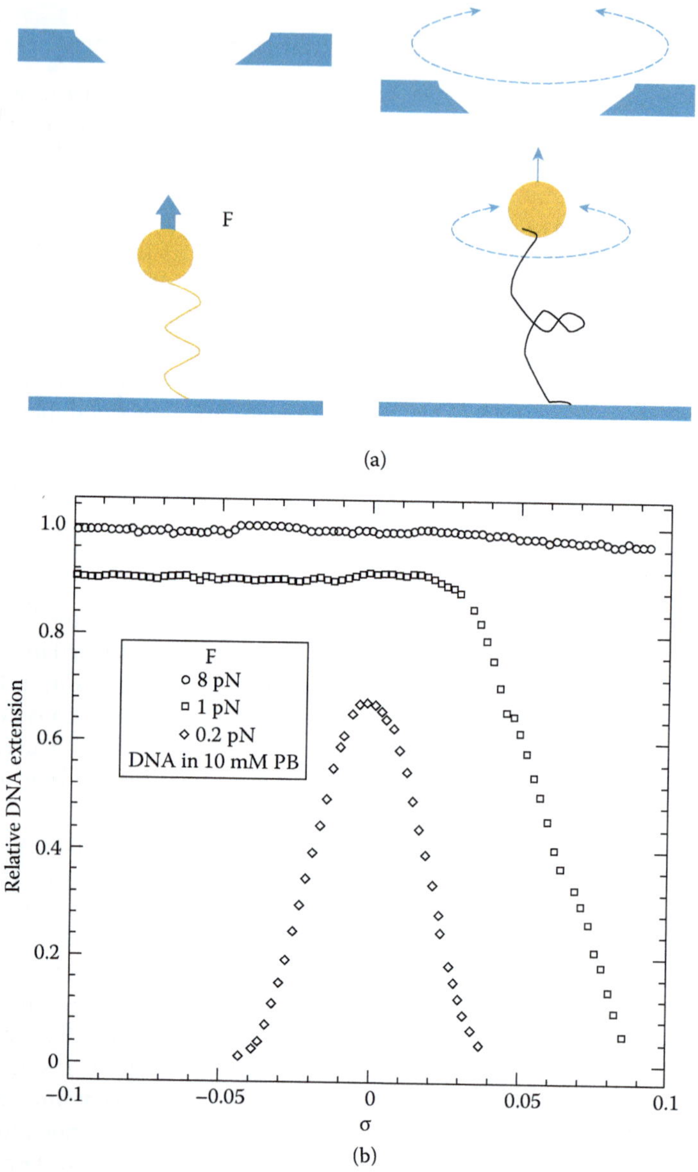

Figure 4.6 Manipulation with DNA by magnetic tweezers. Experimental data for twisting DNA duplex at three different stretching forces in 10 mM phosphate buffer. In the low-force regime (diamonds; $F = 0.4$ pN), the dependence of DNA extension versus calculated DNA supercoiling density appears as a symmetric bell-shaped curve. At intermediate forces (squares; $F = 1$ pN), the dependence is asymmetric. Data for the high force (circles; $F = 8$ pN) do not reveal any dependence of the DNA extension depending on the magnet rotation. (Reprinted with permission from Strick, T. R., et al., *Biophys. J.*, 74, 2016, 1998.)

accompanied by the DNA twisting, so DNA behaves as a torsionless polymer. The interplay between DNA secondary and tertiary structures and mechanics is a complex phenomenon and depends on various factors such as ionic strength, as investigated by Strick et al. (1998).

4.3 Probing of Multiple Samples

One important and unique property of magnetic tweezers is their capability to simultaneously manipulate and monitor a large number of molecules. This property is explained by the fact that the magnetic force generated by the same magnet is roughly uniform over a large area (~100 μm × 100 μm; Figure 4.7). As a result, the rotation of several DNA molecules with individual magnetic beads is possible (De Vlaminck and Dekker 2012). Even with a ~10 μm space between adjacent DNA molecules to prevent crossing of the adjacent molecules with the same length, more than 100 individual molecules can be simultaneously probed.

This capability was demonstrated in a study where the simultaneous manipulation and tracking of 34 DNA-tethered beads at 60 Hz with approximately 1.5 nm resolution and with approximately 10% variation in applied force was achieved (Ribeck and Saleh 2008). Although the multiple manipulation mode is attractive, one should keep in mind that wide-field detection leads to the use of low-resolution lenses, so the resolution is lower. This issue was addressed initially by the authors who developed the approach; however, the lateral resolution as high as 1.5 nm was obtained (Ribeck and Saleh 2008). Further improvement in immobilization approaches and data acquisition allowed the simultaneous probing of as many as 450 beads, yielding statistical data on the mechanics of DNA for 357 molecules from a single experimental run (De Vlaminck et al. 2011). The methodology is very attractive as it allows one to perform kilo-molecule force spectroscopy experiments, opening the possibility of studying rare events in DNA-protein interactions and obtaining statistically significant results from individual experimental runs (De Vlaminck and Dekker 2012).

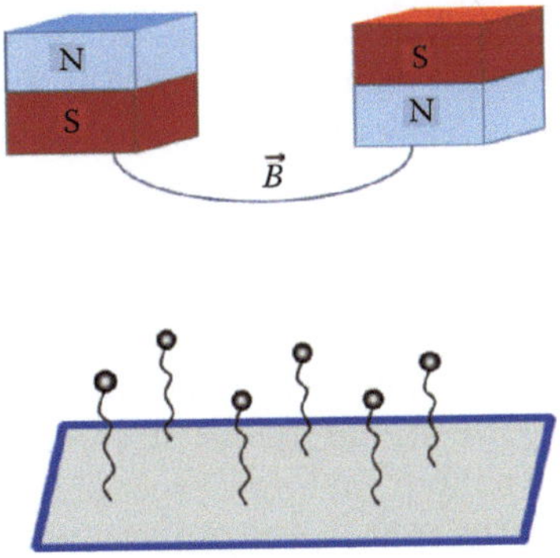

Figure 4.7 Single molecule magnetic tweezers setup for probing of multiple samples.

4.4 Integration of Magnetic Tweezers and Other Single Molecule Instruments

Magnetic tweezers can be integrated in to a number of other instruments enabling measurements of different characteristics of the systems simultaneously or in parallel. Combination with optical tweezers have been demonstrated in a number of applications, and we refer to the review article by De Vlaminck and Dekker (2012) in which recent advances are described. Among these applications, we notice the advantage of the integration of magnetic and optical tweezers in applications that utilize the capability of magnetic tweezers to manipulate multiple beads as described above.

Much more attention was given to the combination of magnetic tweezers with single molecule fluorescence microscopy utilizing Förster resonance energy transfer (FRET). An example of this has been recently reported using DNA hairpin formation for proof of the concept (Kemmerich et al. 2016), and the results of this paper are discussed in Section 4.5.4.

4.5 Selected Applications of Magnetic Tweezers

The initial applications of magnetic tweezers focused on DNA and proteins that alter its structure, including topoisomerases, DNA helicases, recombinases, and single-strand binding proteins. Over the years the DNA substrates have become more intricate, and recent studies have even used magnetic tweezers to construct substrates *in situ* that were used to provide insight into the role of the *Escherichia coli* DNA helicase, RecG in stalled replication fork regression (Manosas et al. 2013).

4.5.1 Probing Interaction of RecG Helicase with a Replication Fork

To do this, the investigators used a 1200 bp hairpin substrate that mimics a DNA replication fork with parental duplex and two single-stranded arms (Figure 4.8a). During DNA replication and fork stalling, it is conceivable that only one of the arms will has an ssDNA gap. Therefore, to construct gapped substrates, a simple modification was used. Assembly begins under low force conditions with the DNA hairpin fully formed (Figure 4.8a, panel 1). Then, force is applied by the magnetic tweezers to completely unzip the substrate (panel 2). This is followed by the introduction of oligonucleotides 90 bases in length and complementary to either the upper (blue strands) or lower regions (red strands) of the substrate. These oligonucleotides are introduced in separate experiments and allowed to anneal. Once the force is lowered, the hairpin is extruded, leaving a fork with one duplex arm (formed by the annealed oligonucleotide) and an ssDNA gap in either the nascent leading or lagging strands (panel 3). These substrates were tested, and RecG was shown to bind to both, and then proceed to rewind the fork concomitant with displacement of the annealed oligonucleotide. Importantly, determination of substrate specificity for RecG was made by determining the on rate for binding. The results show that the on rate for RecG binding to a fork with a gap in the nascent leading strand occurred 8.3-fold faster than to forks with gaps in

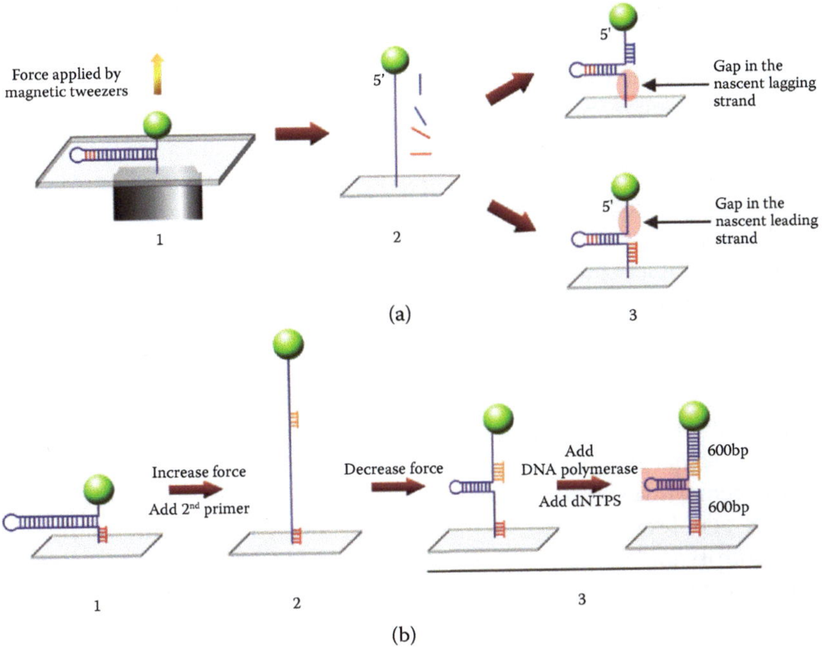

Figure 4.8 Construction of DNA substrates *in situ*. Two types of DNA substrates relevant to fork rescue can be constructed within the confines of the flow cell. (a) Construction of forks with gaps in either the leading or lagging strands. Here, the 1200 bp hairpin is fully stretched by the application of force from the magnetic tweezers. Then, oligonucleotides complementary to the 5′- or 3′-proximal regions are introduced in separate reactions and allowed to bind. Once the opposing force is decreased, a partial hairpin is extruded, and as the oligonucleotide remains annealed to reveal a fork with a gap on the opposite side, either the lagging (top) or leading strand (bottom). (b) Construction of a fork with two duplex arms. As before, the starting point is the 1200 bp hairpin, except now, a short DNA primer is annealed to the 3′ end of the substrate. Then, force is applied to unwind the hairpin, and a second primer is introduced and allowed to bind. When the force is reduced, the DNA length decreases, and a partial hairpin is extruded (~600 bp in length) with primers bound to the opposing arms as indicated. When DNA polymerase is added, the tailed duplex regions are extended from each primer, producing a fork with 600 bp duplex arms. Pink box, parental duplex DNA region of the fork. (Reprinted with permission from Bianco, P.R., *Method*, 108, 40, 2016.)

the nascent lagging strand, a result that is consistent with the substrate specificity determined in bulk phase (Abd Wahab et al. 2013; Bianco 2015, 2016).

In separate experiments, a fork with two duplex arms was required to test whether RecG could regress a fork into a Holliday junction. This fork was constructed *in situ* in the flow cell in sequential steps. First, a primer (red) is annealed to the 1200 bp hairpin (Figure 4.8b, panel 1). Then, force is applied to unzip the substrate, and a second primer (orange) is introduced and allowed to anneal to its target sequence (panel 2). Next, the force is then decreased, allowing the hairpin to partially reform, producing a fork with primers bound to each arm (panel 3, left). Next, T4 DNA polymerase and dNTPs are added, and synthesis ensues using the red and orange primers. Finally, enzyme and unincorporated dNTPs are washed out, revealing a fork with a 600 bp parental duplex region (pink box) and two nascent heteroduplex arms, each 600 bp in length (panel 3, right). Once RecG bound to this fork, it catalyzed fork regression exclusively, which was observed as a shortening of the DNA substrate in Z (Manosas et al. 2013; Bianco 2015, 2016).

4.5.2 Chromatin Assembly Probed by Magnetic Tweezers

In addition, larger and larger protein–DNA complexes have been studied such as nucleosomes (Gupta et al. 2009; Chien and van der Heijden 2014). In these experiments, investigators used a 16.5 kB DNA substrate. Here, both strands of the DNA duplex were attached to the surface via digoxigenin, and at the opposite end, both strands were attached to streptavidin molecules on the surface of a magnetic bead (Figure 4.9a, left panel). Following the introduction of core histone proteins and nucleosome assembly protein 1 (NAP1), nucleosomes formed on the DNA. As nucleosomes are known to wrap DNA, the overall length of the duplex decreased, and this was observed by the drop in molecular extension (ΔZ; Figure 4.9a, right panel). Two versions of the substrate were used. The first used a fully intact duplex, while the second contained a nick in one strand. When histone binding to these substrates was studied, the results were remarkably different (Figure 4.9b and c). DNA binding leading to wrapping of the duplex was more complete with the nicked substrate, as an average of 76 ± 4 nucleosomes assembled resulting in the DNA molecule shortening by 3.79 ± 0.21 μm (Gupta et al. 2009). In contrast, for the topologically constrained molecule, the number of nucleosomes that formed decreased approximately twofold to 39 ± 3, consistent with a change in DNA length by an average of 1.93 ± 0.15 μm (Gupta et al. 2009). The complete assembly occurs only when the positive superhelical tension in the DNA was released by mechanically rotating the DNA.

4.5.3 Repair of the Transcription Stalled Complex by Mfd Protein

Magnetic tweezers have been applied to characterize the mechanism of DNA repair by the Mfd protein that couples transcription and nucleotide excision repair (NER) machineries (Howan et al. 2012). Mfd is a superfamily 2 ATP-dependent DNA translocase. To accomplish the goal, the authors used a 2 kb DNA containing a promoter, a transcribed region, and a terminator that was anchored to the surface by multiple binding sites at one end of DNA and stretched by a 1 μm magnetic bead attached to another end. The force applied to the DNA was kept at 0.3 pN at which DNA can easily form positive or negative supercoils (Figure 4.10a). If DNA is positively supercoiled, the assembly of the complex of RNA polymerase (RNAP) with DNA, is accompanied by the formation of an additional positive four-turns supercoil due to DNA unwinding by

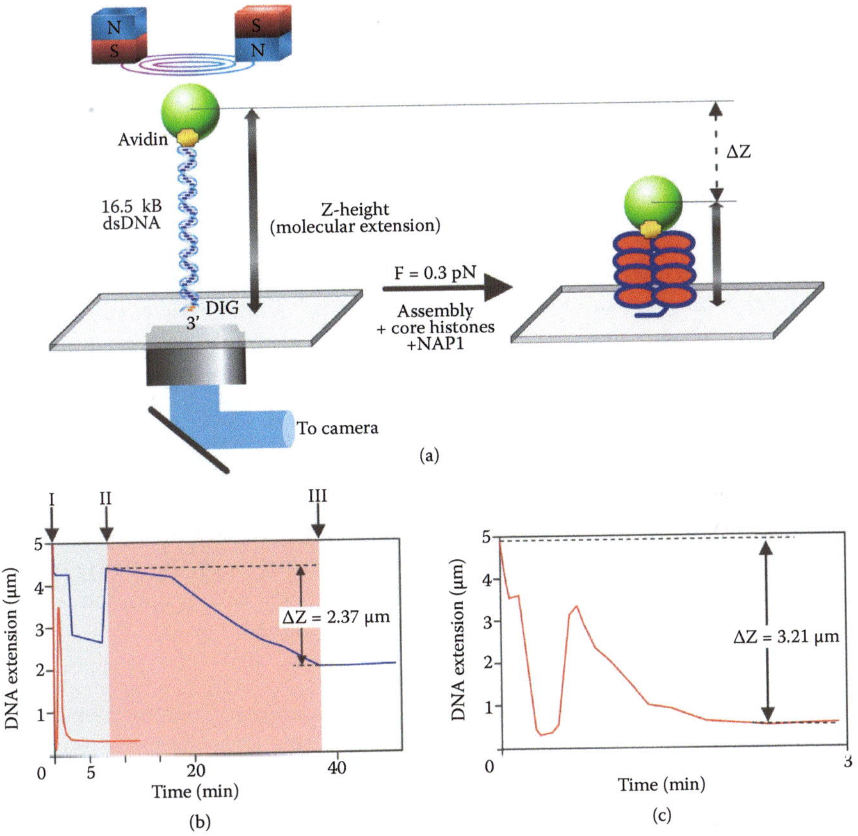

Figure 4.9 Nucleosome assembly is limited by superhelical tension in the DNA. This figure is adapted from reference (Gupta et al. 2009). (a) The experimental setup. In the left panel, the DNA substrate is shown fully stretched by the magnetic tweezers. Following the introduction of core histones and NAP1, nucleosomes form on the fully intact DNA (blue trace) concomitant with DNA wrapping and the associated decrease in molecular extension (ΔZ). (Representative data shown in panels [b] and [c] are from Gupta, P., et al., *Biophys. J.*, 97, 3150–3157, 2009.) (b) The experiment begins with the DNA molecule fully stretched (I). Reaction components are introduced using flow, causing the DNA molecule to fluctuate (gray boxed region). The flow is then turned off (I) and the reaction initiates. The DNA molecule shortens as nucleosomes bind and DNA is wrapped (pink box). The reaction is complete in approximately 40 minutes (III). When the DNA substrate is nicked, nucleosome formation is far more rapid and more complete (red trace). (c), The region from 0 to 3 minutes in panel (b) is expanded to show nucleosome formation on nicked DNA. The rate is more rapid and the extent of reaction is greater as shown by greater ΔZ. (Reprinted with permission from Gupta, P., et al., *Biophys. J.*, 97, 3150–3157, 2009.)

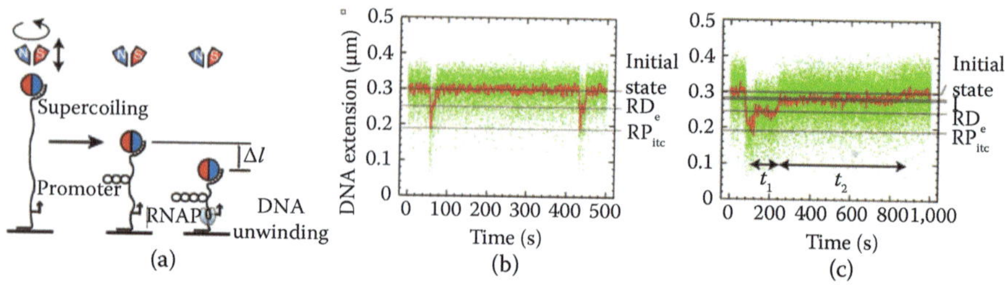

Figure 4.10 Experimental design and real-time dissociation of a stalled RD_e by Mfd. (a) A diagram of DNA stretch and rotating by magnetic tweezers. RNAP-DNA interactions are observed in real time. (b) Single molecule time trace showed there are two complete transcription events. Green points are the raw DNA extension data. Red points are the 1 second average on the raw data. The black line showed three states of DNA extension: initial state, the RNAP/promoter initially transcribing complex (RPitc) during which DNA scrunching occurs, and the ternary elongation complex (RD_e). (c) Representative time trace of DNA extension when 100 mM Mfd was introduced into the buffer. The back process from RD_e to initial state became two distinct steps. The newly occurred long-lived intermediate is formed, which was labeled as I. (Reprinted with permission from Howan, K., et al., *Nature*, 490, 431, 2012.)

RNAP, which is detected by the drop of DNA extension. Therefore, on the time trajectories of the magnetic bead position (Figure 4.10b), the initial position of the bead (~0.3 μm DNA extension value) drops to the RP_{itc} position (~0.2 μm) when RNAP starts the transcription process. Transient state (RD_e, middle black line in Figure 4.10b) corresponds to the promoter escape by the enzyme when the transcription is stalled, either cyclopyrimidine dimers ahead of the transcription complex or by nucleotide starvation. The effect of Mfd is illustrated by the time trajectory data in Figure 4.10c. A new event is detected in Figure 4.10. There is a long-lived state of the complex (t_2) depicted in Figure 4.10c by a thick horizontal green line that is formed after escaping the promoter (after t_1 dwell time). The authors interpret these data suggesting that at the intermediate state, two-thirds of the transcription bubble is rewound and the DNA is bent by approximately 90°C. The Mfd modifies the structure of RD_e state and removes RNAP. Mfd remains bound to DNA a long time to allow the NER system to intervene to repair the lesion.

4.5.4 Integration of Magnetic Tweezers with Single Molecule Fluorescence Microscope

The examples described highlighted one of the major properties of magnetic tweezers—precise and gentle manipulation of the molecular complexes. Most importantly, the external force provided by the magnetic bead to the molecular system can control the state of the system. How do we elucidate local conformations of the molecular complex that can be influenced by external force controlled by magnetic tweezers? This question is especially important if molecular machines are concerned. The necessary information can be retrieved if magnetic tweezers are integrated into such an instrument as a single molecule fluorescence microscope utilizing fluorescence energy transfer (FRET). As described in Chapter 2, single molecule FRET measurements are capable of real-time detection of the dynamics of molecules at the nanometer scale. One of the critical requirements for the instrument combining single molecule FRET with magnetic tweezers is the synchronization of FRET measurements with bead movement defining the external force to the system. This requirement was fulfilled by Kemmerich et al. (2016) in whose study subnanometer magnetic tweezer measurements were correlated with dynamics of DNA constructs provided by the dual-color single molecule fluorescence microscope. The layout of the instrument is shown in Figure 4.11.

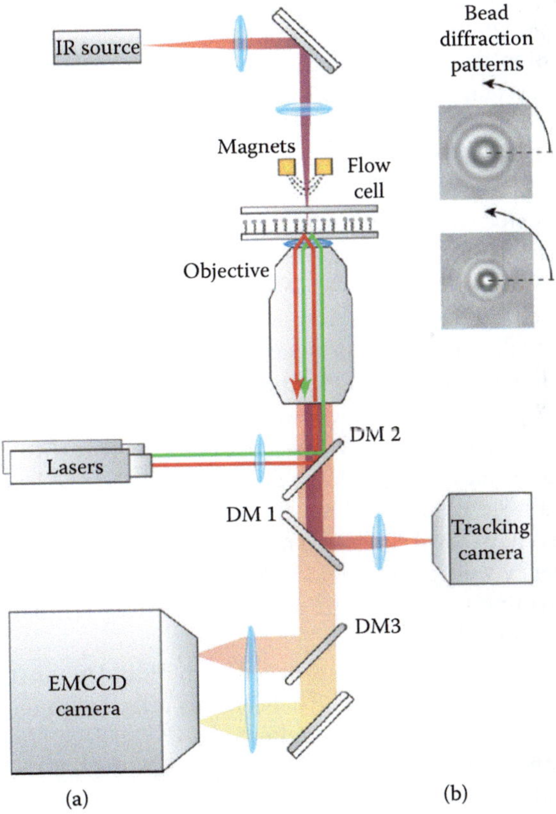

Figure 4.11 The design of magnetic tweezers integrated with molecule fluorescence microscope. (a) A regular magnetic tweezers was integrated with the FRET system. (b) Bead images corresponded to different positions. (Reprinted with permission from Kemmerich, F.E., et al., *Nano Lett.*, 16, 381, 2016.)

The paramagnetic bead with 1 μm diameter is illuminated with a low photon energy infrared (IR) light source (a laser-guided Xe-plasma lamp and the beam is filtered to produce 770–950 nm wavelength) that allowed the authors to track the bead position by the use of tracking cameras (CCD or complementary metal-oxide semiconductor [CMOS] to achieve an acquisition rate up to 2 kHz). The bead diffraction images as illustrated by Figure 4.11b allowed detection of the bead position with respect to the objective lens focus with subnanometer accuracy. Two images correspond to 3 μm difference in the bead positions in vertical direction. The fluorescence is detected by the objective-type total internal reflection fluorescence (TIRF) configuration for single molecule detection with the optics setup for the use of Cy3/Cy5 FRET pair and two lasers for excitation of the dyes (532 and 642 nm). A set of dichroic mirrors (DM1 and DM2) are incorporated to separate florescence light beam from the bead-tracking channel. After separating donor (Cy3) and acceptor (Cy5) signals with DM3 dichroic filter, both fluorescence signals are separated, and each channel is imaged on one-half of the electron-multiplying CCD (EM CCD) camera (data acquisition frequency is 47.5 Hz). Magnets were mounted just above the flow cell and were capable of precise manipulation by the bead (stretching of the tether and rotation of the bead). This system was used to probe two structural transitions in DNA: the equilibrium folding-unfolding process of the 40 bp DNA hairpin, and dynamics of transition of the four-way DNA junction (Holliday junction) between folded and unfolded states. Both processes are controlled by the stretching of DNA provided by an external magnetic field to the bead.

In the case of the hairpin folding-unfolding experiment shown in Figure 4.12a, the hairpin is fully unfolded and DNA is stretched if the force is high. The distance between donor (green) and acceptor (red) was very large, so there was no FRET signal. When low force was applied, the hairpin was fully closed, and the FRET signal was high. In the intermediate regime when a stretching force of 10.6 ± 0.2 pN is applied, the hairpin fluctuates between the open and close states. The extension of DNA was fluctuating between fully open (1.94 μm) and fully close status (1.9 μm); the abrupt change of the length is 37 ± 1.3 nm (Figure 4.12b). Simultaneous with the fluctuating DNA length, the intensities of the two dyes change in a correlated manner, enabling the authors to calculate the FRET efficiency values (Figure 4.12c and d). Both length and FRET values correlate, confirming that under the applied force the DNA hairpin exists in both folded and unfolded states.

Similar experiments were performed with Holliday junctions (Figure 4.13). In these experiments, the magnetic bead was attached to one arm of the junction, and the opposite arm was tethered to the surface. The FRET dyes were attached to the lateral arms of the junction (Figure 4.13a). The Holliday junction can fold, and the distance between the dyes is short generating a strong FRET signal. The distance is large if the junction is in the open conformation, and this state corresponds to a small FRET value due to the large donor-acceptor distance. The stretching force applied to the junction controls the conformational transition, and this is illustrated in Figure 4.13b. At selected stretching force, FRET (Figure 4.13c) also fluctuates periodically, and some of these changes correlate

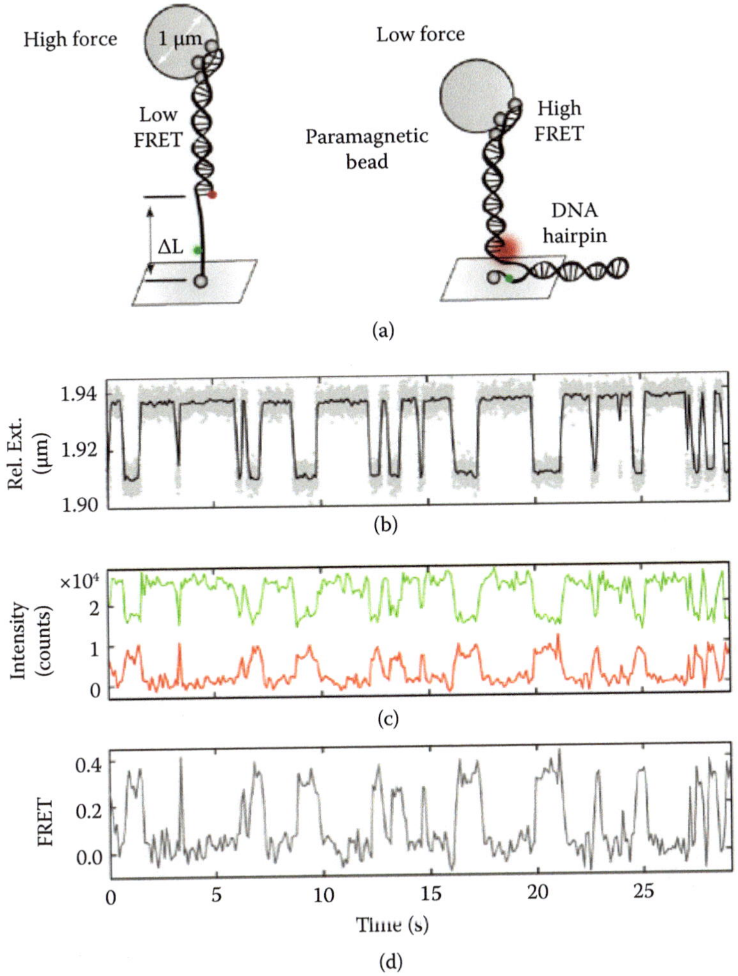

Figure 4.12 Spontaneous opening and closing of a DNA hairpin detected by simultaneous magnetic tweezers and FRET measurements. (a) The DNA sample—a long dsDNA attaching by 40 bp hairpin was stretched by magnetic tweezers. Donor and acceptor fluorophores for the FRET are labeled to the end of ssDNA and shown in green and red, respectively. At high force the hairpin DNA was fully open while it is closed at low force. (b) DNA extension was acquired at high frequency (300 Hz), and the black line showed the average in 10 Hz. (c) The fluorescence intensities over time of donor and acceptor were recorded at 10 Hz. (d) Apparent FRET efficiency calculated from trajectories in (c) was shown. (Reprinted with permission from Kemmerich, F.E., et al., *Nano Lett.*, 16, 381, 2016.)

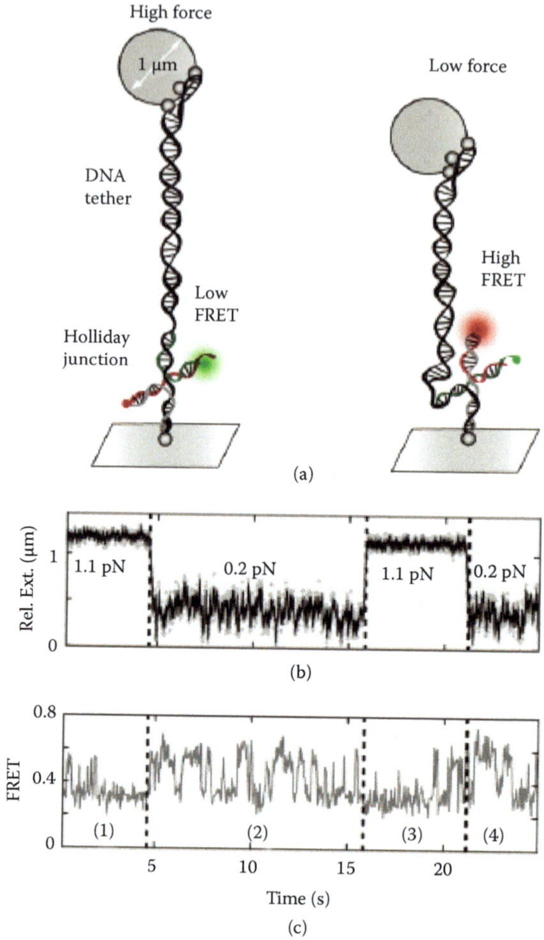

Figure 4.13 Spontaneous opening and closing of a DNA containing a Holliday junction detected by simultaneous magnetic tweezers and FRET measurements. (a) The DNA Holliday junction structure was stretched at two different forces by magnetic tweezers. (b) DNA extension was acquired at high frequency (300 Hz, gray) and filtered by average in 32 Hz (black). (c) Apparent FRET efficiency calculated from trajectories in (b) was shown. (Reprinted with permission from Kemmerich, F.E., et al., *Nano Lett.*, 16, 381, 2016.)

with the fluctuation of FRET; however, in some cases FRET fluctuations occur without changes of the bead position. Note that in these combined measurements, long time trajectories were obtained suggesting that dyes did not bleach significantly due to a weak effect of IR light used for the bead detection. This is an attractive feature of the assembled instrument.

The examples above illustrate only a very few examples of the applications of magnetic tweezers. These are focused on a few systems involving DNA to graphically illustrate major features of magnetic tweezers—precise control of the magnetic bead position and rotation of the bead. However, the applications of the methods are much broader than DNA transactions phenomena. In the following, we briefly mention a few examples. In Min et al. (2015), protein folding/unfolding was probed. Bausch et al. (1999) and Hosu et al. (2003) found that magnetic tweezers were instrumental for the probing of the viscoelastic properties of the cell interior. Note that in Kanger et al. (2008), chromatin was manipulated *in vivo*. Magnetic tweezers have been used in the study of vesicle-substrate adhesion and similarly in studies of cell unbinding forces of the initial adhesion process (Smith et al. 2006; Walter et al. 2006). These reactions are implicated in the infection process. Finally, and as suggested in a recent review, magnetic tweezers will also find utility in studying the infection process directly by measuring forces involved in antibody-antigen interactions and virus to cell adhesion (Zhou and Leake 2016).

4.6 Conclusions

Magnetic tweezers are a widely used single molecule probing technique. The most attractive feature of this technique is the simplicity of the overall design of an instrument that can be built around any type of inverted microscope. In many applications, it rivals the use of optical tweezers instruments described in Chapter 5, although the latter are capable of providing slightly higher spatial resolution. Among the most attractive features of magnetic tweezers, and unique to this approach, is the ability to control torque with a high accuracy and the broad range of the rotation angle. Two advantages that magnetic tweezers have over optical tweezers, are that relatively low-intensity light is used for the illumination of the sample, and the ability to simultaneously probe multiple single molecules. In this chapter, we focused on studies involving DNA, but the range of applications of magnetic tweezers is not limited to these specific systems. Note a publication by Min et al. (2015) in which magnetic tweezers were applied to the probing of protein folding. Recently (Zhou and Leake 2016), studies described the infection process directly by measuring forces involved in antibody-antigen interactions and virus to cell adhesion. As magnetic tweezers instrumentation is in constant development, the range of applications where they find utility is growing.

References

Abd Wahab S, Choi M, Bianco PR (2013) Characterization of the ATPase activity of RecG and RuvAB proteins on model fork structures reveals insight into stalled DNA replication fork repair. *J Biol Chem* 288:26397–26409.

Bausch AR, Moller W, Sackmann E (1999) Measurement of local viscoelasticity and forces in living cells by magnetic tweezers. *Biophys J* 76(1 Pt 1):573–579.

Bianco PR (2015) I came to a fork in the DNA and there was RecG. *Progr Biophys Mol Biol* 117:166–173.

Bianco PR (2016) Stalled replication fork rescue requires a novel DNA helicase. *Methods* (in press). doi: 10.1016/j.ymeth.2016.06.002.

Bliska JB, Cozzarelli NR (1987) Use of site-specific recombination as a probe of DNA structure and metabolism *in vivo*. *J Mol Biol* 194:205–218.

Celedon A, Nodelman IM, Wildt B, Dewan R, Searson P, Wirtz D, Bowman GD, Sun SX (2009) Magnetic tweezers measurement of single molecule torque. *Nano Lett* 9:1720–1725.

Chien FT, van der Heijden T (2014) Characterization of nucleosome unwrapping within chromatin fibers using magnetic tweezers. *Biophys J* 107:373–383.

De Vlaminck I, Dekker C (2012) Recent advances in magnetic tweezers. *Annu Rev Biophys* 41:453–472.

De Vlaminck I, Henighan T, van Loenhout MT, Pfeiffer I, Huijts J, Kerssemakers JW, Katan AJ, van Langen-Suurling A, van der Drift E, Wyman C, Dekker C (2011) Highly parallel magnetic tweezers by targeted DNA tethering. *Nano Lett* 11:5489–5493.

Dulin D, Cui TJ, Cnossen J, Docter MW, Lipfert J, Dekker NH (2015) High Spatiotemporal-Resolution magnetic tweezers: Calibration and applications for DNA dynamics. *Biophys J* 109:2113–2125.

Gore J, Bryant Z, Nollmann M, Le MU, Cozzarelli NR, Bustamante C (2006) DNA overwinds when stretched. *Nature* 442:836–839.

Gosse C, Croquette V (2002) Magnetic tweezers: Micromanipulation and force measurement at the molecular level. *Biophys J* 82:3314–3329.

Gupta P, Zlatanova J, Tomschik M (2009) Nucleosome assembly depends on the torsion in the DNA molecule: A magnetic tweezers study. *Biophys J* 97:3150–3157.

Hosu BG, Jakab K, Bánki P, Tóth FI, Forgacs G (2003) Magnetic tweezers for intracellular applications. *Rev Sci Instrum* 74:4158.

Howan K, Smith AJ, Westblade LF, Joly N, Grange W, Zorman S, Darst SA, Savery NJ, Strick TR (2012) Initiation of transcription-coupled repair characterized at single-molecule resolution. *Nature* 490:431–434.

Kanger JS, Subramaniam V, van Driel R (2008) Intracellular manipulation of chromatin using magnetic nanoparticles. *Chromosome Res* 16:511–522.

Kemmerich FE, Swoboda M, Kauert DJ, Grieb MS, Hahn S, Schwarz FW, Seidel R, Schlierf M (2016) Simultaneous single-Molecule force and fluorescence sampling of DNA nanostructure conformations using magnetic tweezers. *Nano Lett* 16:381–386.

Lionnet T, Allemand JF, Revyakin A, Strick TR, Saleh OA, Bensimon D, Croquette V (2012) Single-molecule studies using magnetic traps. *Cold Spring Harb Protoc* 2012:34–49.

Lipfert J, Hao X, Dekker NH (2009) Quantitative modeling and optimization of magnetic tweezers. *Biophys J* 96:5040–5049.

Lipfert J, Kerssemakers JW, Jager T, Dekker NH (2010) Magnetic torque tweezers: Measuring torsional stiffness in DNA and RecA-DNA filaments. *Nat Meth* 7:977–980.

Lipfert J, Lee M, Ordu O, Kerssemakers JW, Dekker NH (2014) Magnetic tweezers for the measurement of twist and torque. *J Vis Exp* 87:e51503.

Manosas M, Perumal SK, Bianco PR, Ritort F, Benkovic SJ, Croquette V (2013) RecG and UvsW catalyse robust DNA rewinding critical for stalled DNA replication fork rescue. *Nat Comm* 4:2368.

Min D, Jefferson RE, Bowie JU, Yoon TY (2015) Mapping the energy landscape for second-Stage folding of a single membrane protein. *Nat Chem Biol* 11:981–987.

Neuman KC, Nagy A (2008) Single-molecule force spectroscopy: Optical tweezers, magnetic tweezers and atomic force microscopy. *Nat Meth* 5:491–505.

Ribeck N, Saleh OA (2008) Multiplexed single-molecule measurements with magnetic tweezers. *Rev Sci Instrum* 79:094301.

Sinden RR, Carlson JO, Pettijohn DE (1980) Torsional tension in the DNA double helix measured with trimethylpsoralen in living *E.coli* cells: Analogous measurements in insect and human cells. *Cell* 21:773–783.

Smith SB, Finzi L, Bustamante C (1992) Direct mechanical measurements of the elasticity of single DNA molecules by using magnetic beads. *Science* 258:1122–1126.

Smith AS, Lorz BG, Goennenwein S, Sackmann E (2006) Force-controlled equilibria of specific vesicle-substrate adhesion. *Biophys J* 90:L52–L54.

Strick TR, Allemand J-F, Bensimon D, Croquette V (1996) The elasticity of a single supercoiled DNA molecule. *Science* 271:1835–1837.

Strick TR, Allemand JF, Bensimon D, Croquette V (1998) Behavior of supercoiled DNA. *Biophys J* 74:2016–2028.

Tanase M, Biais N, Sheetz M (2007) Magnetic tweezers in cell biology. *Meth Cell Biol* 83:473–493.

Vilfan ID, Lipfert J, Koster DA, Lemay SG, Dekker NH (2009) Magnetic tweezers for single-Molecule experiments. In: *Handbook of Single-Molecule Biophysics* (Hinterdorfer, P., and Oijen, A., eds.), pp 371–395. New York: Springer-Verlag.

Walter N, Selhuber C, Kessler H, Spatz JP (2006) Cellular unbinding forces of initial adhesion processes on nanopatterned surfaces probed with magnetic tweezers. *Nano Lett* 6:398–402.

Zhou Z, Leake MC (2016) Force spectroscopy in studying infection. *Adv Exp Med Biol* 915:307–327.

5 Optical Tweezers

Marco Capitanio
University of Florence
Florence, Italy

Contents

5.1 Introduction

In 1986, when Arthur Ashkin and coworkers first reported about the single-beam gradient force optical trap, now referred to as optical tweezers, they could not imagine that they had invented what now has become one of the most widely used tools in single molecule biology (Ashkin et al. 1986). In the last two decades, optical tweezers have opened up the possibility to manipulate biological molecules and measure protein conformational changes and enzyme kinetics, one molecule at a time. The capacity to apply and measure forces from a few tens of femtonewtons to about 100 piconewtons, which well overlap with the range of forces experienced by biological molecules in their native environment, has elected optical tweezers as an ideal single molecule tool for the study of the mechanical properties of motor proteins and biological polymers, and for investigating regulation of biological systems by mechanical stress. In the last years, force has emerged as a fundamental regulatory factor for biological systems and the conversion of mechanical forces into biochemical signals has been discovered to be at the base of many biological processes fundamental for the development and differentiation of cells (Wozniak and Chen 2009), for their correct function, and for the development of pathologies (Jaalouk and Lammerding 2009). At the molecular scale, force modulates enzymatic activity, induces structural changes in proteins and nucleic acids, alters kinetics of molecular bonds (Marshall et al. 2003; Cecconi et al. 2005), regulates motions of molecular motors (Rief et al. 2000; Reconditi et al. 2004), and has a role in mechanical transduction and sensory functions (Laakso et al. 2008). All these processes are ultimately related to the capacity of force to modulate lifetimes of molecular interactions and transition rates in biochemical reaction cycles that involve motion (Howard 2001). Optical tweezers have been shown to be an ideal tool for the study of these dynamic processes owing to their high spatial and temporal resolutions. In particular, the last few years have seen major improvements in optical tweezers resolution, thus extending the range of measurable quantities and biological systems that can be interrogated with force (Capitanio and Pavone 2013).

Optical tweezers are easily implemented in an optical microscope by focusing a laser beam through the microscope objective to create a large gradient of light intensity. Dielectric microparticles are stably trapped near the beam focus owing to the interaction between the light intensity gradient and the particle itself. A polystyrene or silica microsphere is usually trapped and used as a handle to manipulate single biological molecules bound to it. The bead position is used to probe movements of molecular motors or conformational changes of proteins bound to the bead itself, while the force applied to the biological molecule is simultaneously measured.

In this chapter, we first illustrate how trapping forces are generated, describe how force and displacements are measured, and describe the mechanical behavior of microspheres trapped in optical tweezers. Since trapped microspheres are used as handles to manipulate biological molecules, their mechanical properties influence the outcome of biological measurements. Understanding their behavior when trapped in optical tweezers is thus essential to design and interpret experiments on biomolecules (Section 5.2). We then report in detail some of the prototypical optical tweezers configurations developed to date to investigate single biological molecules, with particular emphasis on recent applications (Section 5.3). We discuss factors affecting spatial and temporal resolutions in the different configurations of measurement and how it is possible to measure subnanometer conformational changes with a microsecond temporal resolution (Sections 5.4 and 5.5). Next, we illustrate principles of instrument design to get a deeper understanding of potentialities and limitations of the technique and its possible implementations (Section 5.6). We finally review a few important applications of optical tweezers to biological problems, exploring combinations of optical tweezers and fluorescence microscopy and outlining applications of optical tweezers in living cells (Section 5.7).

5.2 Principles of Optical Tweezers Trapping

5.2.1 Forces in Optical Tweezers

Forces in optical tweezers arise from the interaction of the laser light with the trapped dielectric particle. Scattering of the trapping light by the dielectric particle alters the photon flux and the momentum carried by the light. For the momentum conservation principle, an equal and opposite momentum change is imparted to the particle. Calculating optical forces in the general case is not a trivial task, but there are two limiting cases for which it becomes easier: the Rayleigh regime, when the particle dimension is much smaller than the wavelength of the trapping laser (Hulst 1981), and the ray optics regimes, for which the wavelength is much smaller than the particle size (Hecht 2002).

In the ray optics approximation, the laser beam can be decomposed into light rays that propagate in straight lines in media of uniform refractive index, carry momentum parallel to the ray direction, and are refracted at the boundary between two media with different refractive index following Snell's law. Figure 5.1a shows how a single light ray is refracted by a microsphere (commonly named *bead*) when the refractive index of the bead is bigger than that of the surrounding medium. The figure also shows how the light ray momentum $\mathbf{p}_{in}$ is changed and the force on the bead $\mathbf{F} = -\mathbf{dp}/\mathbf{dt}$, is generated, as dictated by the momentum conservation principle. The tightly focused beam that constitutes optical tweezers can be decomposed into ray pairs, each with appropriate intensity and direction, as illustrated in Figure 5.1b and c. For each pair, the light rays refracted by the trapped microsphere, the change in momentum, and net force can be drawn. Whatever the position of the bead with respect to the laser beam, the optical force is shown to point toward the beam focus. This qualitative description shows how stable three-dimensional trapping is achieved for this

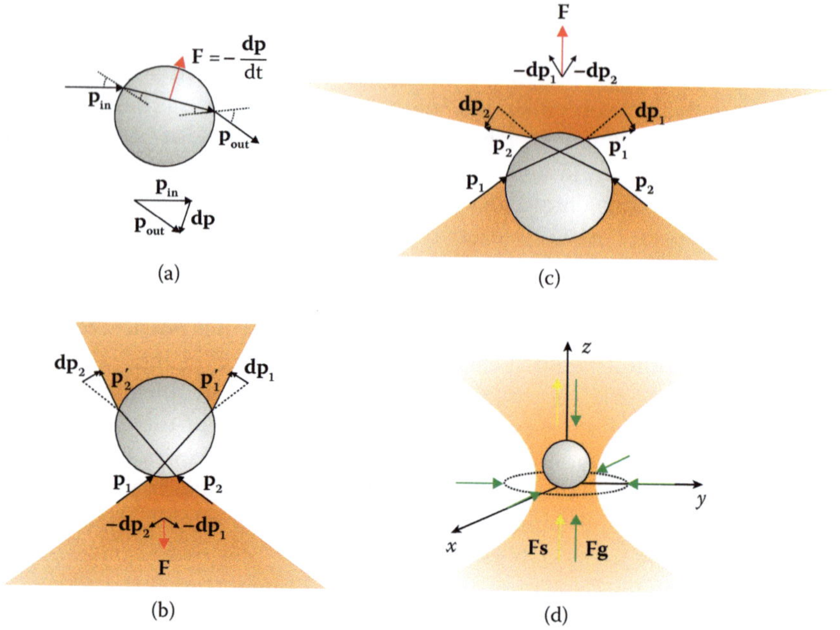

Figure 5.1 (a) A single light ray is refracted two times when it passes through a dielectric sphere. $\mathbf{p}_{in}$ and $\mathbf{p}_{out}$ are the incoming and outcoming momentum, respectively. **dp** is the vector difference $\mathbf{p}_{out} - \mathbf{p}_{in}$. **F** is the optical force applied to the dielectric sphere. The picture represents refraction when the refractive index of the sphere is bigger than that of the surrounding medium. (b and c) A focused trapping laser propagates upward and is refracted by a trapped microsphere positioned above (b) or below (c) the beam focus. The figure shows a pair of light rays symmetric with respect to the beam axis and, for each ray, the refracted ray and the change in momentum. The optical force due to the ray pair, represented in red, moves the bead toward the beam focus, whatever the position of the bead with respect to the laser beam. This picture can be generalized to cases in which the bead is positioned off-axis. (d) Force distribution in optical tweezers. The gradient force (green) points toward the beam focus, whereas the scattering force (yellow) points toward the direction of laser propagation.

geometry. Moreover, trapping forces can be appropriately quantified in the ray optics regime taking into account refraction, reflections, and changes in polarization at dielectric interfaces according to the usual Fresnel formulas (Ashkin 1992). This analysis shows that the force on a dielectric microsphere can be divided into two components: a *scattering force* component pointing in the direction of the incident light and a *gradient force* component pointing in the direction of the gradient of light intensity (Figure 5.1d). The distinction between scattering and gradient forces also emerges naturally in the Rayleigh regime, where the particle acts as a simple point dipole (Neuman and Block 2004), and it is a convenient picture to analyze the trapping stability of optical tweezers. In fact, stable trapping is achieved when the gradient force, which pushes the bead toward the laser focus, predominates over the scattering force, which pushes the bead away in the direction of propagation of the beam. Therefore, the beam convergence angle should be as large as possible in optical tweezers to maximize the light intensity gradient, which is achieved by using large numerical aperture objectives and slightly overfilling the objective back aperture (Neuman and Block 2004).

Several articles, reviews, and books describe in detail the physics of the interaction between light and dielectric particles and accurately quantify forces applied by optical tweezers in the Rayleigh and ray optics regimes, as well as in the intermediate case in which the dimension of the bead is comparable to the light wavelength (Hulst 1981; Ashkin et al. 1986; Ashkin 1992; Wright et al. 1994; Ren et al. 1996; Neuman and Block 2004). A common feature of all regimes is that, for small distances, the trap behaves as a Hookean spring with an elastic constant k, the *trap stiffness*. For the purpose of this book, we therefore describe the force experienced by the bead (F_{trap}) as pointing toward the trap center and growing linearly with the distance of the bead center from the trap center (x): $F_{trap} = -k \cdot x$ (Figure 5.2a). The trap stiffness is constant in the axial plane (perpendicular to laser propagation) and smaller in the longitudinal direction owing to the smaller intensity gradient along that direction (Capitanio et al. 2002). Since in most optical tweezers geometries the force is applied in the axial plane (Section 5.3), in the present text we indicate a direction in this plane as x, unless otherwise specified. Beyond the linear region, a small zone of near-constant force follows at the border of the potential well, after which the force rapidly drops to zero (Figure 5.2b). Optical tweezers stiffness varies with laser power, beam focusing (i.e., the numerical aperture of the objective), bead dimension, and the refractive index of the bead and the surrounding medium. Several articles describe the dependence of tweezers force on those parameters. Table 5.1 shows typical stiffness values as a function of laser power and bead dimensions, which are the two parameters that can be more easily adjusted to match the experimental needs. In particular, since the gradient of the trapping

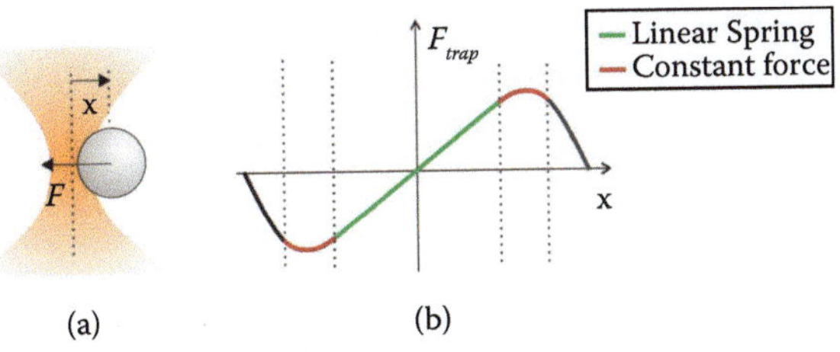

Figure 5.2 Force exerted by optical tweezers. (a) A dielectric microsphere is stably trapped near the laser beam focus. A lateral displacement of the bead (x) is opposed by a restoring force F_{trap}. (b) For small displacements of the bead from the trap center, the force grows linearly with x. Beyond the linear region, a near-constant force region follows, after which the force rapidly drops to zero.

Table 5.1 Trap Stiffness (k) and Position Detector Conversion Factor (β) as a Function of Laser Power and Bead Diameter

Laser Power (mW)	Bead Diameter (μm)	k (pN/nm)	β (nm/V)
3.2	1	0.024	50
8		0.05	
16		0.091	
32		0.18	
48		0.27	
3.2	0.5	0.01	270
	1	0.024	50
	2.6	0.004	125
	5	0.002	330

Source: Capitanio M, et al. *Review Sci Instrum* 73:1687–1696, 2002.

light intensity is proportional to the light intensity itself, tweezers stiffness grows proportionally to laser power. On the contrary, the dependence of stiffness on bead diameter is nonlinear and shows a maximum for bead dimensions close to the trapping light wavelength (Capitanio et al. 2002). Box 5.1 describes common methods to measure the bead position x, quantify the tweezers stiffness k, and, thus, measure forces and displacements in optical tweezers.

Box 5.1 Measuring Forces and Displacements in Optical Tweezers

Most of optical tweezers' success comes from their capacity to precisely measure the size of protein conformational changes and quantify forces applied by enzymes or biological polymers. Movements produced by biological molecules can be measured from the displacement of the attached bead through the use of position photodetectors (see Section 5.6.3). To this end, however, the photodetector voltage signal (V) must be first converted into a position signal (x). Analogously, to measure forces (F_{trap}), we need to know how the force applied on the bead is related to the measured bead displacement (x). In other words, we need two calibration curves $x(V)$ and $F_{trap}(x)$ to measure displacements and forces in optical tweezers. Since both calibration curves are linear for typical bead displacements (±500 nm using a 1 μm diameter bead), optical tweezers calibration reduces to the acquisition of two calibration parameters, the *position detector conversion factor* β (or the *detector sensitivity* $\alpha = 1/\beta$), and the optical tweezers *stiffness* k (Capitanio et al. 2002):

$$x = \beta \cdot V \tag{5.1}$$

$$F_{trap} = -k \cdot x \tag{5.2}$$

As mentioned in Section 5.2.1, optical tweezers stiffness changes with laser power, beam focusing, bead dimension, and the refractive index of the bead and the surrounding medium. The detector sensitivity mainly depends on the detection strategy adopted, on the photodetector and its electronics (which is discussed in Section 5.6.3), as well as on the bead properties. Similar to the tweezers stiffness, the detector sensitivity is maximal for bead dimensions close to the trapping light wavelength. Table 5.1 shows typical detector sensitivity as a function of bead diameter. Several strategies have been developed for optical tweezers calibration, each one with peculiar advantages and drawbacks. Several articles and reviews describe the different calibration methods and how calibration parameters change with the bead and the trapping laser properties (Ghislain et al. 1994; Gittes and Schmidt 1998b; Capitanio et al. 2002). Free downloadable software for optical tweezers calibration is also available (Hansen et al. 2006; Osterman 2010).

Optical tweezers calibration is most commonly based on the analysis of thermal fluctuations of a trapped bead, which are discussed in detail in Section 5.2.2.2. For example, by applying the equipartition theorem as in Equation 5.9, we can obtain the trap stiffness by simply measuring the variance of the bead position:

$$k = \frac{k_B T}{\langle x^2 \rangle}, \tag{5.3}$$

where k_B is the Boltzmann constant, and T is the absolute temperature. However, to this end, we should be able to measure bead displacements, which is usually not the case unless the position detector conversion factor β has been, somehow, previously obtained. Frequency analysis of the thermal fluctuations of the trapped bead can be used to get both k and β. The power spectrum of the position signal V can be obtained from the power spectrum of bead thermal fluctuations $S_x(f)$, as described in Equation 5.11, and from Equation 5.1,

$$S_V(f) = \frac{1}{\beta^2} S_x(f) = \frac{1}{\beta^2} \frac{k_B T}{\pi^2 \gamma \left(f_C^2 + f^2\right)} \tag{5.4}$$

where γ is the hydrodynamic drag coefficient, and $f_c = k/2\pi\gamma$ is the cut-off frequency (see Section 5.2.2.2). The β^2 term derives from the fact that the power spectrum is the squared Fourier transform of the bead position noise. The two calibration parameters k and β can thus be easily obtained by fitting Equation 5.4 to the experimental power spectrum of the position detector signal V recorded from a trapped bead, leaving k and β as free parameters (Figure 5.3).

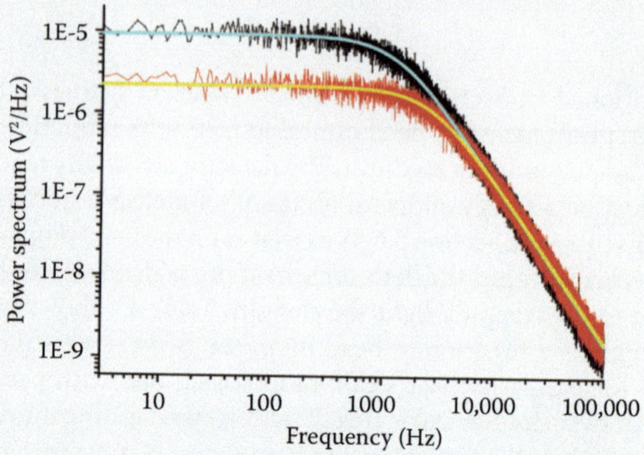

Figure 5.3 Power spectra of a 510 nm diameter bead trapped in optical tweezers. Each trace is the average of 30 power spectra obtained from 1 second recording acquired at 200 kHz. Data are fitted with Equation 5.4; from the fit, we obtain a cut-off frequency *fc* of 1273 and 2703 Hz for the black and red curves, respectively, which correspond to trap stiffnesses of 0.04 pN/nm and 0.08 pN/nm, respectively. The nm/V conversion factor β is 75 nm/V from both calibrations. (From Capitanio M, et al. *Review Sci Instrum* 73:1687–1696, 2002.)

5.2.2 Dynamics of a Trapped Bead in Optical Tweezers

Understanding how a bead moves when trapped in optical tweezers is fundamental to decipher experiments performed on single biological molecules, measure the amplitude and dynamics of protein conformational changes, optimize temporal and spatial resolution, and calibrate position detector sensitivity and trap stiffness (see Box 5.1). The simple mechanical model of a bead trapped in optical tweezers that we develop in this chapter is later extended to more complex configurations, including attached proteins and multiple optical tweezers. Here, we limit our analysis to the motion of the bead in one direction (x); extension to two or three dimensions can be easily derived.

The mechanical system considered here is represented in Figure 5.4a. The trapped microsphere is regarded as a point-like mass m, which is subjected to external forces from the optical tweezers, represented as a spring, and from a viscous buffer in which it is immersed, represented as a dashpot.

As introduced in the preceding paragraph, optical tweezers can be regarded as a Hookean spring, which exerts a force directly proportional to the spring extension x:

$$F_{\text{trap}} = -kx \tag{5.5}$$

where k is the *trap stiffness*.

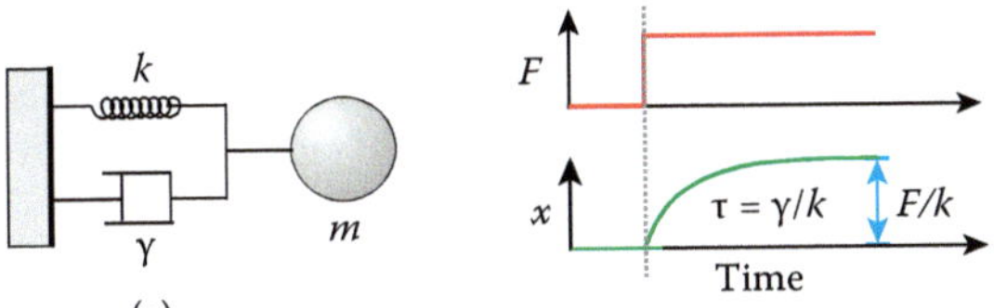

Figure 5.4 (a) Mechanical model of a microsphere of mass m, trapped in optical tweezers with stiffness k, and immersed in a viscous solution with drag coefficient γ. (b) Dynamics of the microsphere after a force step F. The bead exponentially moves from $x = 0$ to the new equilibrium position $x = F/k$ with a time constant $\tau = \gamma/k$. Inertial terms are neglected in the approximation of low Reynolds numbers.

An object that is moving inside a viscous fluid with velocity v experiences a force opposing its motion. For small objects, such as the micrometer-sized bead considered here, cells, or proteins in water solution, the viscous force is always predominant over the inertial force (i.e., the object mass m times its acceleration). The ratio between inertial and viscous force is called the Reynolds number (Re), and Re < 1 for small objects. Small Reynolds numbers imply that the fluid flow is laminar and the viscous force is proportional to velocity (Batchelor 1967; Howard 2001), as stated by Stokes' law:

$$F_{drag} = -\gamma v \tag{5.6}$$

where γ is the *hydrodynamic drag coefficient*. The drag coefficient depends on the shape of the object and on the fluid viscosity (η). For a sphere of radius R far from any surface, we get $\gamma = 6\pi\eta R$ (Batchelor 1967).

The equation of motion of the bead in one direction (x) can be obtained after applying Newton's second law:

$$\gamma \frac{\partial x}{\partial t} + kx = F \tag{5.7}$$

where F is the additional external force acting on the bead in x-direction, and we have neglected the inertial term $m = \frac{\partial^2 x}{\partial t^2}$ in the approximation of low Reynolds numbers.

5.2.2.1 Motion After a Perturbation

We first describe the motion of the trapped bead after a perturbation by an external force. This can happen, for example, when a protein attached to the bead undergoes a conformational change. Next, we see that the bead motion depends on both trap stiffness and viscous damping on the bead.

Let's assume that the bead is at rest at time 0 in the center of the trap (i.e., $x(0) = 0$, where the spring is relaxed), and a constant external force F puts the system out of equilibrium. At time 0, the bead starts moving at constant velocity $\frac{\partial x}{\partial t} = \frac{F}{\gamma}$, as described by Equation 5.7. The spring then elongates in the direction of the force, applying an increasing opposing force, and the bead velocity consequently decreases, until it drops to zero. The system, thus, reaches a new equilibrium position in which the elastic force equals the external force: $kx = F$. The time course of this motion can be obtained by solving the differential Equation 5.7. We get:

$$x(t) = \frac{F}{k}\left[1 - \exp\left(-\frac{t}{\tau}\right)\right]; \tau = \frac{\gamma}{k} \tag{5.8}$$

This motion is represented in Figure 5.4b. The time constant τ is named the *relaxation time* and is a fundamental characteristic of the system, which tells us the time required by the trapped bead to reach the new equilibrium position. The relaxation time sets a limit in the time resolution of optical tweezers. It is not possible to detect protein dynamics faster than the relaxation time, simply because the bead dynamics occur with a time constant that is dictated by the relaxation time. Table 5.2 shows typical relaxation times of trapped beads in typical optical tweezers configurations, some of which are described in Section 5.3. In the case considered here (single-bead configuration in Table 5.2), we get $\tau \sim 0.1$ ms for a 1 μm diameter bead trapped in optical tweezers with typical stiffness (0.1 pN/nm). We should, however, point out that the relaxation time changes significantly when the bead is tethered to the coverslip surface through an attached protein, which can increase significantly the system stiffness. A 1 μm bead bound to a protein with typical stiffness (k = 1 pN/nm), shows a relaxation time of about 10 μs (see Table 5.2, Single bead + kinesin). We discuss this point in detail in Section 5.5, when we analyze the temporal resolution of optical tweezers in typical experimental configurations.

Table 5.2 Relaxation Time versus Stiffness and Bead Diameter for Different Configurations of Measurement in Optical Tweezers

System	Bead Diameter (μm)	System Stiffness (pN/nm)	Relaxation Time (μs)
Single bead	1	0.1	94
	0.2		19
Rigid dumbbell	2 × 1	2 × 0.025	377
	2 × 0.5		188
Single bead + kinesin	1	1	9.4
	0.2		1.9
Rigid dumbbell + myosin	2 × 1	1	19
	2 × 0.5		9.4

5.2.2.2 Thermal Fluctuations of a Trapped Bead

In the previous paragraph, we examined the dynamics of a trapped bead subjected to an external force *F*. However, we did not take into account thermal forces, which strongly affect the motion of micrometer-sized beads immersed in water solution at room temperature. We now analyze the motion of the trapped bead subjected to thermal forces only.

Under this assumption, *F* in Equation 5.7 represents the fluctuating force originated by random collisions of water molecules with the trapped microsphere. Because of the stochastic nature of the resulting motion, the bead position can be analyzed in statistical terms only—that is, by measuring the probability $p(x)$ of finding the bead in position x or by calculating average position values such as the mean position $\langle x \rangle$ or the mean square position $\langle x^2 \rangle$.

The trapped bead randomly oscillates about position $x = 0$, which is the bead equilibrium position where the spring is relaxed (see Figure 5.5a, left). Therefore, $\langle x \rangle = 0$, as can be easily inferred from the symmetry of Equation 5.7 with respect to the equilibrium position. In this case, the mean square position $\langle x^2 \rangle$ equals the position noise variance $\sigma_x^2 = \langle (x - \langle x \rangle)^2 \rangle$. This quantity is an important parameter that quantifies how much the trapped bead oscillates because of thermal forces. The equipartition theorem gives us a simple way to calculate $\langle x^2 \rangle$ (Landau et al. 1980). For an object in a harmonic potential $U(x) = \frac{1}{2}kx^2$:

$$\frac{1}{2}k_BT = \frac{1}{2}k\langle x^2 \rangle \;\Rightarrow\; \langle x^2 \rangle = \sigma_x^2 = \frac{k_BT}{k} \tag{5.9}$$

where k_B is the Boltzmann constant and T the absolute temperature. Equation 5.9 is telling us that bead thermal fluctuations increase with temperature and decrease when restrained by a trap with higher stiffness, a quite intuitive result represented in Figure 5.5a.

At thermal equilibrium, we can also derive $p(x)$ from Boltzmann distribution (Landau et al. 1980; Howard 2001):

$$p(x) = \frac{1}{Z}\exp\left[-\frac{U(x)}{k_BT}\right] = \frac{1}{Z}\exp\left(-\frac{x^2}{2\frac{k_BT}{k}}\right) \tag{5.10}$$

where $Z = \int_{-\infty}^{+\infty} \exp\left[-U(x)/k_BT\right]dx$ is a normalizing constant, called the partition function, which assures that the sum of all probabilities adds up to one (Landau et al. 1980). Equation 5.10 tells us that thermal noise is Gaussian distributed, with $\langle x \rangle = 0$ (Figure 5.5a, right). From Equation 5.10, we also find that the variance of the Gaussian distribution is $\sigma_x^2 = \frac{k_BT}{k}$, which confirms Equation 5.9 obtained from the equipartition theorem.

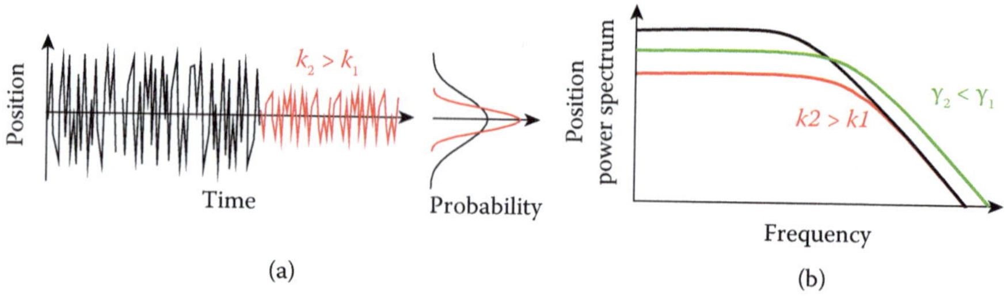

Figure 5.5 (a) Position fluctuations of a trapped bead owing to thermal noise. *Left*: Amplitude of position fluctuations depends on the trap stiffness according to Equation 5.9. *Right*: Position noise is Gaussian distributed, as described by Equation 5.10. (b) Frequency distribution of position noise represented in a log–log scale. From Equation 5.11, position fluctuations decrease with stiffness k at low frequency ($f << f_C \Rightarrow S_x(f) = 4k_BT\gamma/k^2$), but not at high frequency ($f >> f_C \Rightarrow S_x(f) = k_BT/\pi^2\gamma f^2$), as represented by the black and red traces. Position fluctuations decrease with the drag coefficient γ for low frequency, but conversely, they increase for lower values of γ at high frequency, as represented by the black and green lines.

Equations 5.9 and 5.10 fully describe thermal noise amplitude at equilibrium. However, it is also important to know that thermal noise amplitude varies depending on the time (or frequency) scale considered, and largely decreases at high frequencies. This has important implications when studying dynamic processes occurring at different time scales, as shown later. The power spectrum of the bead position (i.e., the squared magnitude of its Fourier transform) describes how noise amplitude changes with frequency (Gittes and Schmidt 1998a; Neuman and Nagy 2008):

$$S_x(f) = \frac{k_BT}{\pi^2\gamma\left(f_C^2 + f^2\right)} \tag{5.11}$$

where $f_C = k/2\pi\gamma = 1/2\pi\tau$ is the *cut-off frequency* of the motion, and it is inversely proportional to the relaxation time of the trap defined in Equation 5.8. The position variance σ_x^2 in a given frequency range can be simply obtained as the area (integral) under the power spectrum in that frequency range (Landau et al. 1980; Neuman and Nagy 2008). Equation 5.11 implies that noise amplitude is constant for frequencies well below the cut-off frequency ($f << f_C$) and decreases as $1/f^2$ for high frequencies ($f >> f_C$), as depicted in Figure 5.5b (in the log–log scale of the graph, $1/f^2$ becomes a line with slope –2). The position variance σ_x^2 over all frequencies can be obtained as the total area (integral) under the power spectrum described by Equation 5.11, which gives us again Equation 5.9: $\sigma_x^2 = k_BT/k$.

The presence of the cut-off frequency in thermal noise is a consequence of the relaxation time of bead motion. In fact, since bead movements cannot occur at time scales shorter than the relaxation time, collisions with water molecules occurring at these short time scales are not transformed into significant bead movements, and thermal noise is consequently reduced.

We also observe that both trap stiffness k and bead dimension R ($\gamma = 6\pi\eta R$) affect the position noise amplitude through the cut-off frequency of motion. A higher stiffness determines a decrease of the noise amplitude at low frequencies. On the contrary, smaller beads cause a reduction of the noise amplitude at low frequencies, but an increase at high frequencies, so that the total noise in Equation 5.9 is not dependent on bead dimension (Figure 5.5b).

In the more general case in which the bead is subjected to both thermal forces and a constant external force, the overall motion is a directed motion described in the previous paragraph by Equation 5.8 superimposed to the random motion described in the present paragraph.

5.3 Configurations of Measurement

5.3.1 Single-Bead Geometry

In the simplest optical tweezers geometry, named *single-bead* or *single-trap geometry*, optical tweezers are kept at a fixed position and a trapped bead monitors conformational changes and movements of a protein, which is linked on one end to the bead and on the other end to the coverslip surface (Figure 5.6a). This configuration has been largely used to investigate processive molecular motors such as conventional kinesin (Svoboda et al. 1993), which can move continuously

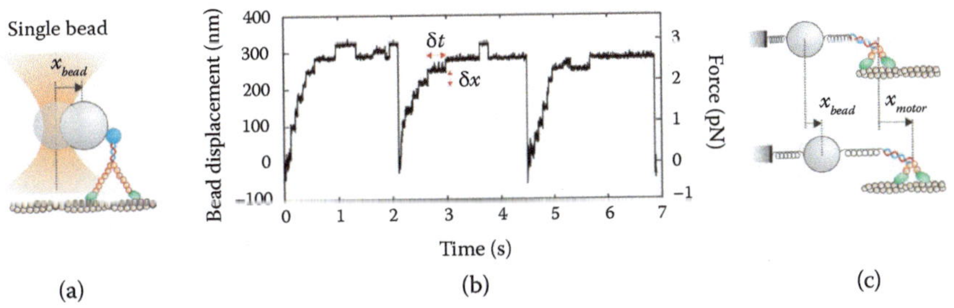

Figure 5.6 Single-bead or single-trap geometry. (a) The drawing represents a myosin V molecule attached to a trapped bead and proceeding along an actin filament stuck on the coverslip surface. The optical trap position is fixed and the bead displacement x_{bead} measures protein displacement. (b) An example showing stepwise movement of a single myosin Va motor in a single-bead assay. Consecutive 36 nm stepwise movements (δx) and dwell times between steps (δt) are clearly visible. Backward steps are also visible at high force. The force was calculated from the displacement of the bead from the trap center times trap stiffness (0.009 pN/nm; right axis). Stall force is about 2.5 pN. (Adapted from Uemura, S., et al., *Nat. Struct. Mol. Biol.*, 11, 877–883, 2004. With permission.) (c) Measured bead displacements (x_{bead}) do not correspond to motor displacements (x_{motor}) but depend on the motor protein displacement and on the values of the trap and protein stiffness.

along a microtubule for up to several microns, or myosin Va (Uemura et al. 2004), which proceeds along actin filaments. In this configuration, the force applied to the processive motor increases as the protein displaces the bead from the trap center while translocating along its track ($F_{trap} = -k_{trap}x_{bead}$; in the following, we specify stiffness k and displacements x with appropriate subscripts to avoid ambiguity). When the maximum force that the motor protein can develop is reached, the motor stalls (Figure 5.6b). Stall forces of kinesin (Svoboda and Block 1994), RNA polymerase (Yin et al. 1995), and many other motors have been measured using the single-bead geometry.

Moreover, various analysis approaches have been developed to detect the single steps from a position recording of a stepping motor (Carter et al. 2008). From such analyses, it is possible to measure the amplitude of the motor working stroke (the step δx in Figure 5.6b), as well as the distribution of dwell times between steps (δt), which gives us precious information on the kinetics of the mechanochemical cycle of the motor protein (Howard 2001).

Care must be taken when measuring molecular movements in an optical trap, since bead movements (x_{bead}) do not always correspond to the movements of the biological molecule (x_{motor}). In fact, the system is compliant owing to the finite stiffness of the trap (k_{trap}) and the molecule and linkages connecting the molecule to the bead and to the coverslip surface (k_{motor}) (Figure 5.6c). When the motor takes a step, both springs are stretched, so that $x_{bead} \neq x_{motor}$. Referring to Figure 5.6c,

$$x_{bead} = x_{motor}\frac{k_{motor}}{k_{motor} + k_{trap}} \tag{5.12}$$

so that $x_{bead} = x_{motor}$ only when $k_{trap} << k_{motor}$. In general, it is important to evaluate both trap and motor stiffness and use Equation 5.12 to accurately determine the motor working stroke.

5.3.2 Two-Bead Geometry

Double optical tweezers (two-bead, double-trap, or dumbbell assay) or a single trap together with a micropipette have been widely employed to investigate the mechanical properties (force–extension curves) of single dsDNA and ssDNA molecules (Smith et al. 1996), as well as single RNA molecules (Liphardt et al. 2001). In this configuration, one of the two traps is held at a fixed position and probes force applied to the polymer, while the second trap or the micropipette is displaced to stretch the polymer and measure its force–extension curve (Figure 5.7a, top). The force–extension curve, which is a straight line for Hookean linear springs, can display complex behavior in biopolymers such as DNA (Figure 5.7a, bottom). Different regimes in the force–extension curve allow the deciphering of different conformational states of the polymer and changes in molecular structure (van Mameren et al. 2009).

More recently, the two-bead geometry has been applied to investigate protein folding by using dsDNA handles to tether the protein between the two beads (Figure 5.7b, top) (Cecconi et al. 2005; Gebhardt et al. 2010; Stigler et al. 2011). The force–extension curve of a single protein transitioning between folded and unfolded states can reveal intermediates conformations, interactions between

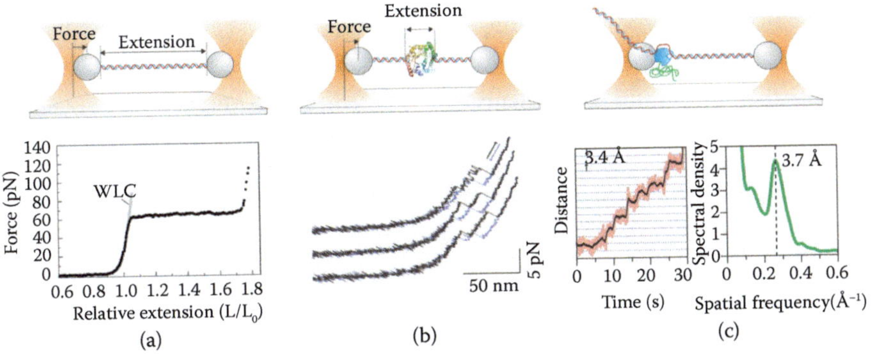

Figure 5.7 Two-bead or double-trap assay. (a) Measurement of force–extension curve of a biopolymer. *Top*: The left trap is stationary and measures the force applied to the polymer. The right bead moves in steps or ramps and, for each displacement, the force applied to the polymer and its extension are measured. *Bottom*: Typical force–extension curve of a 3′-3′ attached DNA, with free 5′ ends. The elastic properties of DNA below the overstretching force of 65 pN are well described by the extensible worm-like chain (WLC) model (*gray line*). At 65 pN, the DNA molecule undergoes the overstretching transition, during which the intrinsic contour length of the DNA increases from 100% to about 170%. (Adapted from van Mameren, J., et al., *Proc. Natl. Acad. Sci. U S A*, 106, 18231–18236, 2009. With permission.) (b) Probing unfolding–refolding of single proteins by force. *Top*: Analogous to panel (a), the left trap measures force while the right trap is moved to change protein extension. *Bottom*: Folding and unfolding of single calmodulin stretch-and-relax cycles for velocity v = 500 nm/s. (Adapted from Stigler, J., et al., *Science*, 334, 512–516, 2011. With permission.) (c) Dynamics of DNA-processing enzymes. *Top*: Cartoon represents a single, transcriptionally active molecule of RNA polymerase (RNAP, blue) attached to a bead held in a trap and tethered via the upstream DNA to another trapped bead. During elongation, the DNA tether lengthens and the beads move apart. *Bottom Left*: A representative record for a single RNAP molecule transcribing under 18 pN of assisting load, median-filtered at 50 ms (pink) and 750 ms (black). Horizontal lines (dotted) are spaced at 3.4 Å intervals. *Bottom Right*: The power spectrum of the average autocorrelation function derived from position histograms shows a peak at the dominant spatial frequency, corresponding to the inverse of the fundamental step size, 3.7 ± 0.6 Å. (Adapted from Abbondanzieri, E.A., et al., *Nature*, 438, 460–465, 2005. With permission.)

protein domains, and energy landscape, which provides a conceptual framework for understanding protein folding (Figure 5.7b, bottom).

The same geometry has been largely used to monitor DNA and RNA processing enzymes, such as the bacteriophage ϕ29 (Smith et al. 2001b; Chemla et al. 2005), T7 DNA polymerase (Wuite et al. 2000), RNA polymerases (Abbondanzieri et al. 2005; Fazal et al. 2015), helicases (Dumont et al. 2006; Garcia-Garcia et al. 2015), and the ribosome (Wen et al. 2008; Kaiser et al. 2011) (Figure 5.7c). In those experiments, the molecular motor is usually attached to one of the trapped beads and is linked to the second bead through a DNA or RNA tether. The processive movements of the motor along the nucleic acid tether shorten or lengthen it, so that the motor movements can be detected through bead movements (Figure 5.7c, top). When properly optimized, this assay allows resolving single base-pair stepping of RNA polymerase (Abbondanzieri et al. 2005), as illustrated in detail in Section 5.4, where we discuss spatial resolution in optical tweezers (Figure 5.7c, bottom). Moreover, in this assay, enzyme processivity can be efficiently analyzed under constant force by using force-clamp techniques that we describe in Section 5.3.4.

5.3.3 Three-Bead Assay

In the three-bead assay, a biopolymer is suspended between two optically trapped beads, and a binding protein is attached to a third bead stuck onto the coverslip (Figure 5.8a). This configuration has been extensively used to investigate nonprocessive molecular motors such as skeletal, cardiac, and smooth muscle myosin (Finer et al. 1994; Molloy et al. 1995; Capitanio et al. 2006). Nonprocessive molecular motors interact with their cytoskeletal filament for a single enzymatic cycle during which they produce a conformational change (working stroke) and unbind.

Position recordings from a single skeletal muscle myosin interacting with actin appear as in Figure 5.8b (Capitanio et al. 2006). When myosin is not bound to actin, the dumbbell oscillates within the traps because of thermal motion, the amplitude of the oscillations depending on the overall system stiffness, as described by Equation 5.9. When myosin binds to actin, the system stiffness increases, and position noise consequently decreases. Acto-myosin interactions

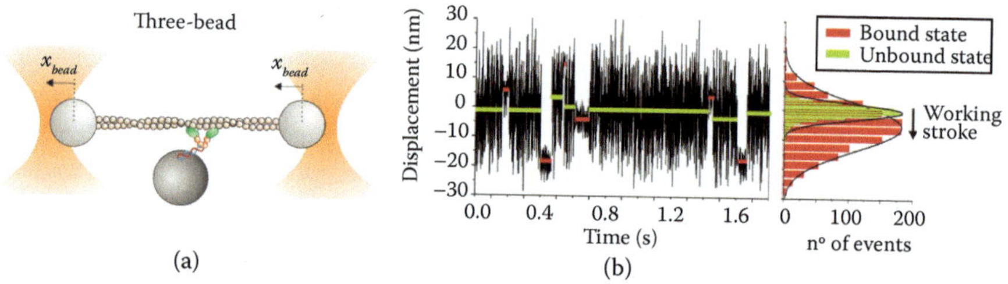

Figure 5.8 Three-bead assay. (a) Cartoon illustrates an actin filament connected to polystyrene beads held by optical tweezers. A single myosin molecule is attached onto a third bead stuck to the coverslip surface. Movements of the actin filament produced by the attached myosin are measured through bead displacements (x_{bead}) (b) *Left*: Position recording while myosin is interacting with actin. Red and green lines indicate the average position of bound and unbound events, respectively. *Right*: Distributions of the average position of bound and unbound events of a 100 s position recording containing several hundreds of interactions. The working stroke is obtained from the displacement between the centers of the two distributions. (Adapted from Capitanio, M., et al., *Micr. Res. Tech.*, 65, 194–204, 2004. With permission.)

are thus detected from noise reduction of the bead position signal (red lines in Figure 5.8b). Various methods have been developed for the separation of low-variance events (bound state) from high-variance events (unbound state) (Knight et al. 2001; Smith et al. 2001a). Since myosin displaces the actin filament very rapidly after binding (~1 ms) and interactions last for several tens of milliseconds at the low ATP concentrations used in these experiments, the position of bound events corresponds here to the position after the working stoke is completed. Therefore, myosin working stroke can be evaluated from the displacement between the average position of bound and unbound events (Figure 5.8b, right). Moreover, the distribution of event duration describes the kinetics of the mechanochemical cycle of the myosin motor.

By replacing the actin filament with a microtubule, the three-bead assay has been adapted to study microtubule-based molecular motors such as cytoplasmic dynein (Walter et al. 2010, 2012) or ncd kinesin (deCastro et al. 2000). Alternatively, the same assay has been used to study DNA binding proteins (Capitanio et al. 2012) and processing enzymes (Skinner et al. 2004). The same assay was recently applied to investigate actin-binding proteins involved in mechanotransduction processes (Ren et al. 2009; Buckley et al. 2014). Some of the experiments reported here required dynamic trap configurations that are described in the next paragraphs.

5.3.4 Force Clamp

A more sophisticated geometry is the force clamp or isotonic clamp, which has been developed in the single-bead, two-bead, or three-bead geometries, using different strategies. In active force clamps, the distance between the bead and the trap (x_{bead}) is continuously monitored through a position detector (Visscher et al. 1999; Rief et al. 2000). As the distance changes owing, for example, to the movements of a motor protein, a feedback system rapidly moves the trap to keep the distance, and thus the force, constant ($x_{bead} = const. \Rightarrow F_{trap} = -k_{trap}x_{bead} = const.$) (Figure 5.9a, top). Active force clamps were extensively applied in a single-bead geometry to study processive molecular motors such as kinesin (Visscher et al. 1999), myosin Va (Rief et al. 2000), myosin VI (Rock et al. 2001), and RNA polymerase (Wang et al. 1998). In these experiments, the bead and the trap move together and probe protein movements (Figure 5.9a, bottom trace). Usual position detection methods, such as the back focal plane detection described in Section 5.6.3, directly measure the distance of the bead from the trap center (x_{bead}). Once the trap stiffness is known, the position detector is used to measure the force ($F_{trap} = -k_{trap}x_{bead}$), which is kept constant by the feedback loop (besides high-frequency thermal noise that cannot be compensated by the feedback because of its limited frequency response) and the feedback signal, i.e., the trap position (x_{trap}), measures protein movements.

A double-trap configuration, in which one trap is stationary and the other one is force-clamped, has been applied to study nucleic acids or protein folding (Liphardt et al. 2001; Gebhardt et al. 2010; Bianco et al. 2011; Rognoni et al. 2012) and movements of DNA and RNA processing enzymes (Kaiser et al. 2011; Fazal et al. 2015). Sometimes the stationary trap is replaced by a micropipette that holds one of the beads in a fixed position. Figure 5.9b shows how such a configuration

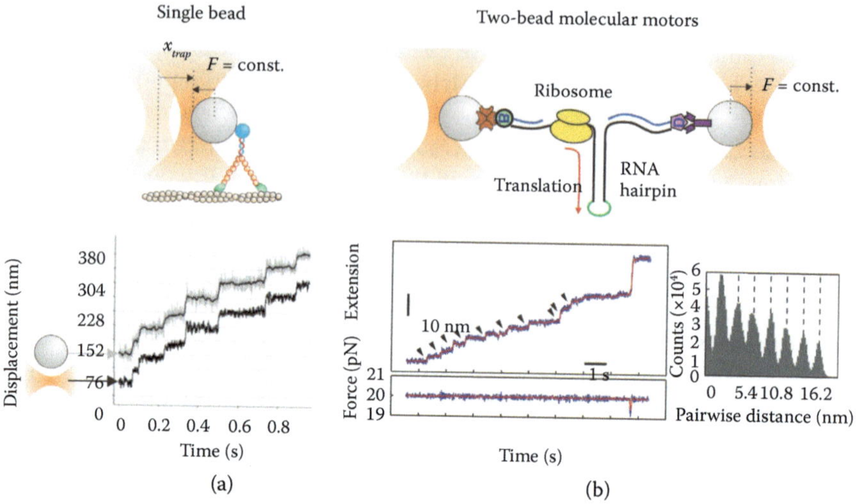

Figure 5.9 Force-clamp geometries. (a) Single bead. *Top*: The drawing represents a myosin-V molecule attached to a trapped bead, which proceeds along an actin filament stuck on the coverslip surface. A feedback system moves the trap to keep force on the bead and on the motor protein constant. Trap displacements (x_{trap}) measure protein displacements. *Bottom*: Example record of the bead position (gray curve) and trap position (lower black curve) as the myosin-V molecule steps along the actin filament. The distance between the two is kept constant by the feedback loop; thus, the myosin-V molecule is always kept under constant load. The thin black line within the gray curve is a filtered bead position signal (box filter, 15 ms). (Adapted from Rief, M., et al., *Proc. Natl. Acad. Sci. USA*, 97, 9482–9486, 2000. With permission.) (b) Two-bead force clamp applied to molecular motors. *Top*: Cartoon illustrating the experimental configuration used to study translation by a single ribosome. (Adapted from Wen, J.D., et al., *Nature*, 452, 598–603, 2008. With permission.) A single ribosome was stalled at the 59 side of the mRNA hairpin construct, which was then held between two polystyrene beads. The right bead applied a constant force below the force required to unfold the hairpin, and the extension change was monitored in time from the movements of the force-clamped trap. The static trap was substituted by a micropipette depending on the RNA construct used for translation. *Bottom left*: Extension and force trajectories during translation. The data were collected at 200 Hz (blue traces) and smoothed to 10 Hz (red). Discrete steps are indicated by arrowheads. The 18 nm rip at the end of the record corresponds to spontaneous opening of the remaining approximately 18 bp hairpin ahead of the translating ribosome. *Bottom right*: Pairwise distance analysis of the extension trajectory in the first 10 s of the record, after correction for drift, shows codon-by-codon stepping.

(Continued)

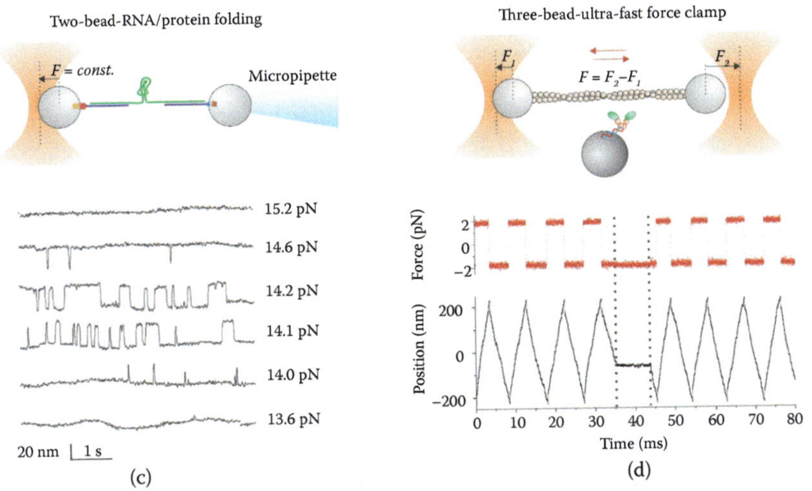

Figure 5.9 (continued) Force-clamp geometries. (c) Two-bead force clamp applied to RNA and protein folding. *Top*: The experimental configuration used to study reversible unfolding of a single RNA molecule under force (Liphardt et al. 2001). Individual RNA molecules (green) were attached to polystyrene beads by RNA/DNA hybrid "handles" (purple). One bead was held in a force-clamped optical trap, and the other bead was held by a micropipette. *Bottom*: Length versus time traces of the RNA hairpin at various constant forces. The end-to-end distance of the RNA hairpin hopped back and forth by 18 nm, signaling the repeated folding and unfolding of a single RNA molecule. By increasing the pre-set force, it was possible to tilt the folded-to-unfolded equilibrium toward the unfolded state (Liphardt et al. 2001). (d) Three-bead ultrafast force clamp. *Top*: Schematic of the operational principle illustrating constant force $F = F_2 - F_1$ applied to the dumbbell through two feedback systems clamping the force on the left bead to $-F_1$ and on the right bead to F_2. *Bottom*: Force and position of the dumbbell. The net force is switched between $+F$ and $-F$ ($F = 2$pN) to keep the dumbbell within a confined spatial interval (±200 nm). When myosin binds to actin, the force is transferred to myosin and the dumbbell stops (vertical dotted lines).

has been used to follow translation by a single ribosome, one codon at a time (Wen et al. 2008). Figure 5.9c depicts the experiment by Liphardt et al. in which they used mechanical force to induce the unfolding and refolding of a single RNA hairpin (Liphardt et al. 2001). The RNA hairpin could be mechanically unfolded and when kept at constant force within a critical force range, hopped between folded and unfolded states (Figure 5.9c, bottom trace).

A three-bead force clamp (named ultrafast force-clamp spectroscopy) was recently developed to study intermittently binding proteins such as nonprocessive molecular motors and transcription factors binding DNA under constant force (Capitanio et al. 2012) (Figure 5.9d). Here, a constant force is applied to the dumbbell through a double feedback system. The force is alternated back and forth so that the dumbbell oscillates in a triangular wave fashion when it is free in solution. When the molecule binds the filament, the force is rapidly transferred to the molecule, and the dumbbell stops its movement (Figure 5.9d, bottom trace). Molecular interactions occur under the constant force imposed by the force clamp on the dumbbell. This technique has several advantages in terms of temporal resolution and is discussed in detail in Section 5.5.3, in relation with methods that allow pushing temporal resolution of optical tweezers to the limit.

Despite the higher complexity of implementing force clamps, single molecule data collected in this configuration allow a direct measurement of force dependence of conformational transition rates, thus enabling the reconstruction of the energy landscapes of molecular structural states. Force-clamp geometries also open up the possibility of measuring the mechanical output of molecular motors, such as the working stroke and the average distance traveled before dissociating (run length), as a function of load. This configuration also eliminates the complication of correcting the measured distances taking into account the compliances of the motor protein and its linkages with the bead and the coverslip surface (Equation 5.12). In fact, all elastic elements remain at a fixed stretching length under force clamp ($x_{bead} = F_{trap}/k_{trap} = const.$), and motor movements are precisely measured from the feedback signal ($x_{motor} = x_{trap}$).

As we see in Section 5.6, piezo or motorized mirrors can be used to steer the laser beam and move the trap, although much faster response (μs) can be reached using acousto-optic deflectors (AODs) (Lang et al. 2002) or electro-optic deflectors (EODs) (Valentine et al. 2008). The limited bandwidth of the feedback loop can sometimes be restrictive (Elms et al. 2012), depending on the time constant of the process under study relative to the response time of the feedback loop. Alternative configurations have been developed to overcome such limitation. For example, Nambiar et al. created a one-dimensional region of constant force extending over several micrometers by rapidly line scanning the trapping light while simultaneously modulating its intensity (Nambiar et al. 2004). Greenleaf et al. developed a passive all-optical force clamp exploiting the constant-force region of optical tweezers near the border of the potential well (Figure 5.2b) (Greenleaf et al. 2005). Such a clamp was successfully employed to measure the force-velocity relationship for transcription by RNA polymerase (Abbondanzieri et al. 2005) (Figure 5.7c), as well as folding trajectories of nucleic acids under constant force (Woodside et al. 2006; Greenleaf et al. 2008).

5.3.5 Position Clamp

Opposite to force clamp, where force on the protein is kept constant and protein movements are measured, the position clamp (also called the isometric clamp) prevents protein movements and measures the force that is developed. Molecular isometric clamp was first developed by Finer et al. to measure forces exerted by single actomyosin complexes in the three-bead geometry (Finer et al. 1994) and subsequently refined by Takagi et al. (2006). In this latter implementation, one of the two beads, termed the *transducer bead*, detects movements of the dumbbell, whereas the force on the other bead, the motor bead, is changed using an AOD to oppose the detected movements and maintain the bead-actin-bead assembly at its initial position (Figure 5.10a). Actin–myosin interactions are thus detected

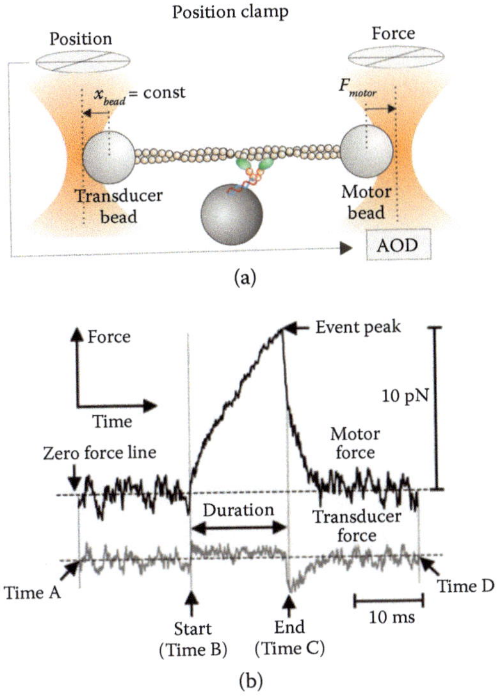

Figure 5.10 Position or isometric clamp. (a) The left bead detects movements of the dumbbell (x_{bead}), whereas the right bead moves using an AOD to oppose the detected movements. The right bead measures the force applied by the motor protein (F_{motor}). (b) A single acto-myosin interaction in the position clamp. After the force peak the force rapidly declines. The time between the start (Time B) and the point at which the rate of force declines is fastest (Time C) is defined as the duration of an episode. (From Takagi, Y., et al., *Biophys. J.*, 90, 1295–1307, 2006. With permission.)

from the increase in force exerted on the motor bead (Figure 5.10b). Myosin isometric force measured from these experiments was ~9 pN, close to that estimated from high-resolution fiber mechanics studies (Piazzesi et al. 2002). The same assay recently allowed the measurement of force-dependent interaction between actin and cardiac myosin (Greenberg et al. 2014) and different isoforms of myosin I (Laakso et al. 2008, 2010; Greenberg et al. 2012).

5.3.6 Dynamic Force Spectroscopy

Another strategy to probe the effect of force on molecular bonds is dynamic force spectroscopy (DFS), in which the distribution of rupture forces of molecular bonds is measured at different loading rates (Evans 2001) (Figure 5.11). Constant loading rates can be applied by moving the trapped bead at constant velocity ($dF_{trap}/dt = -k_{trap} \cdot [dx_{bead}/dt] = -k_{trap} \cdot v_{bead}$) (Arya et al. 2005; Dame et al. 2006; Guo and Guilford 2006; Bianco et al. 2007; Lewalle et al. 2008; Pyrpassopoulos et al. 2010), or by clamping the position of the bead relative to the optical trap ($x_{bead} = const.$)

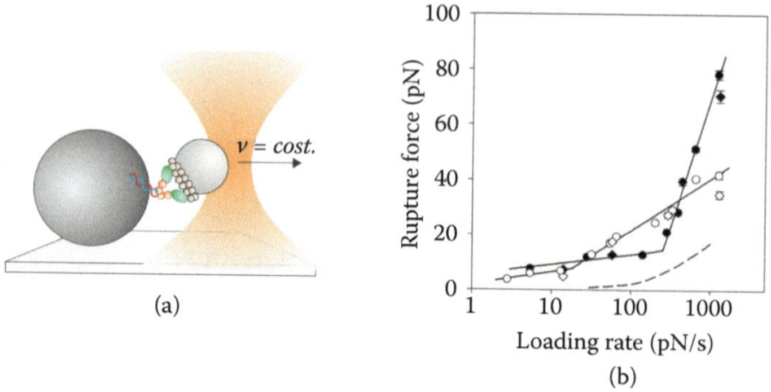

Figure 5.11 Dynamic force spectroscopy. (a) Cartoon representing dynamic force spectroscopy applied to actin–myosin bond. The molecular bond is subjected to constant loading rates and rupture forces and bond lifetimes are measured. (b) The characteristic rupture force of actomyosin bonds over a range of loading rates. At higher loading rates, actomyosin–ADP bonds (full circle) appear to be stronger than rigor bonds (empty circles). Corresponding data from single-headed HMM in each state (full and empty diamonds) are superimposed. The dashed line indicates the rigor behavior predicted from step-load data (From Guo, B., et al., *Proc Natl Acad Sci U S A*, 103, 9844–9849, 2006.). Error bars represent standard error of the mean (SEM).

and increasing the optical power at a constant rate ($dF_{trap}/dt = -(dk_{trap}/dt) \cdot x_{bead}$) (de Messieres et al. 2012). Data collected with DFS can be converted, using appropriate models, to get the kinetics of bond detachment under constant forces (Evans 2001; Dudko et al. 2008). Such an approach has been commonly used for weak bonds, in which the kinetics are too fast to effectively work in force-clamp geometries. However, a proper conversion relies on the accuracy of the model assumptions. Moreover, recently, alternative techniques such as ultrafast force-clamp spectroscopy opened the possibility to direct probing rapid force-dependent kinetics of molecular bonds, with unbinding rates on the order of several tens of kHz.

Among the many reports, DFS was applied to study the interaction between actin and skeletal muscle myosin (Guo and Guilford 2006; Lewalle et al. 2008) or titin (Bianco et al. 2007), and between lipid bilayers and myosin I (Pyrpassopoulos et al. 2010). The binding of the glycoprotein Ibα (GPIbα) subunit to the A1 domain of von Willebrand factor (VWF), which mediates the formation of platelet plugs for arterioles and is associated with the bleeding disorder von Willebrand disease, was recently studied by DFS (Arya et al. 2005; Kim et al. 2010).

5.4 Spatial Resolution

As described in the previous paragraphs, one of the key features of optical tweezers is the capability to measure conformational changes and displacements produced by single biological molecules. Such movements range from several nanometers for cytoskeletal molecular motors, down to one base pair (~3.6 angstrom) for DNA and RNA processing enzymes (Greenleaf et al. 2007; Veigel and Schmidt 2011). High spatial resolution is, thus, a fundamental requirement for the study of a wide range of biological molecules. Molecular displacements can be measured through different position detection systems (Section 5.6.3). The most sensitive detectors to date, which are based on interferometry (Svoboda et al. 1993; Allersma et al. 1998; Gittes and Schmidt 1998c), allow detection of angstrom movements on a time scale of 1 ms or better. When properly optimized, position detectors do not set a lower limit on spatial and temporal resolution of optical tweezers. We showed in Section 5.2.2.2 that the movements of a bead trapped in optical tweezers are governed by Brownian fluctuations owing to thermal forces, which are inherent in single molecule experiments performed in liquid solution at ambient temperature. Thermal noise, indeed, sets fundamental limits on position and force measurements with single molecules.

5.4.1 Thermal Noise

Understanding which parameters affect thermal noise is crucial to optimizing spatial resolution in position measurements with optical tweezers. We have already showed that the amplitude and frequency distribution of thermal fluctuations of a trapped bead depend on the stiffness and viscosity of the system (Section 5.2.2.2). Here, we consider the single-bead geometry described in Section 5.3.1, in which the trapped bead is linked to a motor protein moving along its cytoskeletal filament, and we evaluate how to optimize the measurement of the motor step size. The system has a combined stiffness k, which comprises the trap stiffness (k_{trap}) and the stiffness of the molecule and the linkages connecting the molecule to the bead and to the coverslip surface (k_{motor}). The drag coefficient is mainly due to the bead, since the viscous drag of the motor protein is negligible with respect to that of a micrometer-sized bead (Figure 5.12a). The power spectrum of thermal motion is again given by Equation 5.11, where $\gamma = \gamma_{bead}$ and $k = k_{trap} + k_{motor}$ (both the optical trap and the motor protein are rigidly coupled to the ground, so that the two springs act in parallel and sum with each other, as represented in Figure 5.12b). The amplitude of thermal noise (position variance) is given by Equation 5.9. In an actual experiment, however, the bead position is always measured over a finite frequency interval or bandwidth Δf, owing, for example, to the cut-off frequency of the position detector, which sets a high frequency limit, and the limited time the motor protein spends in a fixed position, which sets a low frequency limit. In such a case, the position variance is the area under the power spectrum within the interval Δf (e.g., the turquoise area depicted in Figure 5.13). Therefore, an easy way to reduce the measured thermal noise is by shrinking the measurement bandwidth Δf. This can be achieved, for example, by simply filtering data with a moving time-average window. Such a procedure allows bringing bandwidth in the low-frequency regime ($f << fc$) in which Equation 5.11 reduces to $S_x(f << f_C) = 4k_BT\gamma_{bead}/k^2$ and

$$\sigma_x = \sqrt{S_x(f << f_C)\cdot \Delta f} = \frac{\sqrt{4k_BT\gamma_{bead}\Delta f}}{k} \tag{5.13}$$

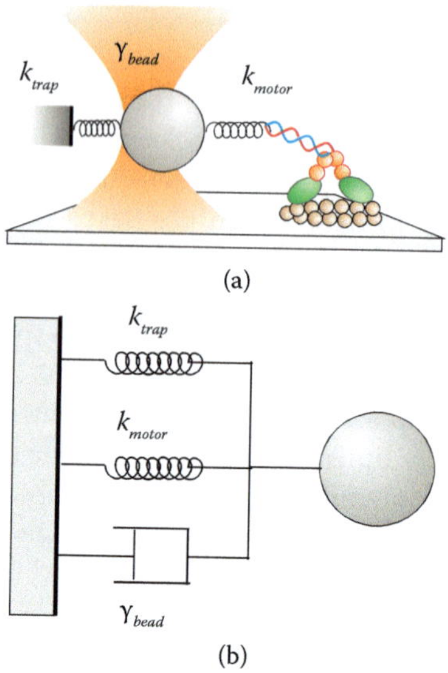

Figure 5.12 Mechanical elements affecting thermal noise in a single-bead assay. (a) k_{trap} is the trap stiffness and k_{motor} comprises the motor protein stiffness in series with the stiffness of the linkages connecting the protein to the bead and the coverslip surface. γ_{bead} is the bead viscous drag coefficient. (b) Mechanical model of the single-bead assay represented in (a); k_{trap}, k_{motor}, and the viscous drag γ_{bead} act in parallel on the microsphere.

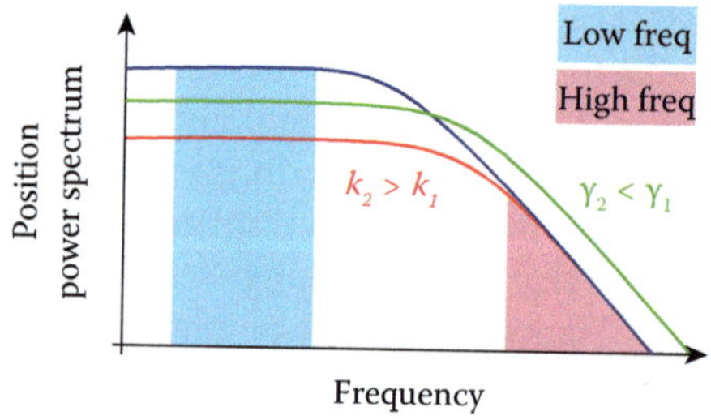

Figure 5.13 Power density spectrum of thermal noise of the bead position. When the stiffness of the system increases ($k_2 > k_1$), the noise amplitude decreases at low frequencies ($f \ll f_C$) (turquoise area), but it is unchanged at high frequencies (pink area). When the viscous drag decreases ($\gamma_2 < \gamma_1$), the low-frequency noise decreases, whereas the high-frequency noise increases. The area under the power spectrum stays constant.

Equation 5.13 tells us that, besides reducing measurement bandwidth, low position noise can be achieved by increasing system stiffness (large k) and using small beads (small γ_{bead}). In fact, the total amplitude of thermal noise is reduced when the bead is linked to a stiffer element (Figure 5.13, red curve). On the contrary, a smaller bead does not change the total amplitude of thermal motion (Equation 5.9 is independent of γ) but changes its spectral distribution, leaving less noise at the low frequencies in which measurements are usually made (Figure 5.13, green curve).

The system stiffness can be increased either by using stiffer traps (increasing laser power) or by using stiffer molecules, or both. However, increasing trap stiffness has a zero net effect on the signal-to-noise ratio (SNR) of the measurement of molecular motor step size. In fact, combining Equations 5.12 and 5.13, the SNR is given by

$$\mathrm{SNR} = \frac{x_{bead}}{\sigma_x} = \frac{k_{motor} x_{motor}}{\sqrt{4 k_B T \gamma_{bead} \Delta f}} \tag{5.14}$$

We see from Equations 5.12 through 5.14 that increasing trap stiffness indeed reduces thermal noise, but it also reduces movements of the probe relative to the real movements of the molecule, so that the SNR is not changed. Equation 5.14 is still valid in case we are measuring motor step size using an ideal force clamp; in fact, since the optical force is kept constant ($dF_{trap} = -k_{trap} dx_{bead} = 0$), the thermal force is not influenced by the trap stiffness but only by the molecule stiffness; moreover, in a force clamp, the bead and trap movements are equal to the motor movement (Gittes and Schmidt 1998a).

Since high trap stiffness does not help in measuring steps of single molecular motors, researchers have worked on increasing the stiffness of biological constructs. This can be achieved by using functional truncations of the molecule under study (the stiffness of an elastic element decreases with its length [Howard 2001]), using rigid molecules to connect the protein under study to the bead and the coverslip surface (e.g., using biotin–avidin or covalent bonds), and by pretensioning the molecule. In fact, the stiffness of biological molecules is usually nonlinear and increases with the applied tension. However, even after pretensioning, typical protein stiffness does not exceed a few pN/nm.

Unfortunately, bead dimension also cannot be indefinitely reduced owing to a consequent reduction in optical tweezers force and sensitivity in position measurement (see Table 5.1 and Box 5.1), so that it is usually very difficult to work with microsphere diameters below 200 nm.

Similar conclusions can be drawn for the two-bead assay, in which a single molecule is tethered between two microbeads trapped in optical tweezers (Figure 5.7). However, trap and molecular stiffness sum up differently in this case. For example, with reference to Figure 5.14, the stiffness of the molecule (k_{DNA}) and the stiffness of the right trap (k_2) act in series, and this combined elastic element acts on the left bead in parallel with the left trap (k_1). This is analogous to the single bead described above, where the rigid attachment to the coverslip is replaced by the right bead trapped in optical tweezers. As a consequence, the added Brownian noise due to the second microsphere deteriorates the SNR, which now depends on the stiffness of the second trap (k_2) and is always smaller compared to the single-bead assay (Moffitt et al. 2006). However, Moffitt et al. demonstrated that by detecting the positions of both trapped microspheres ($x1_{bead}$ and $x2_{bead}$), correlations in their motions can be exploited to maximize the SNR. In fact, the difference coordinate ($x_d = x1_{bead} - x2_{bead}$), with appropriate choice of experimental parameters displays a SNR that is always superior to that of the single-trap geometry (Moffitt et al. 2006).

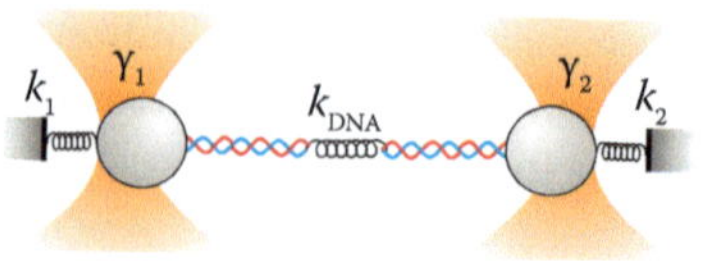

(a)

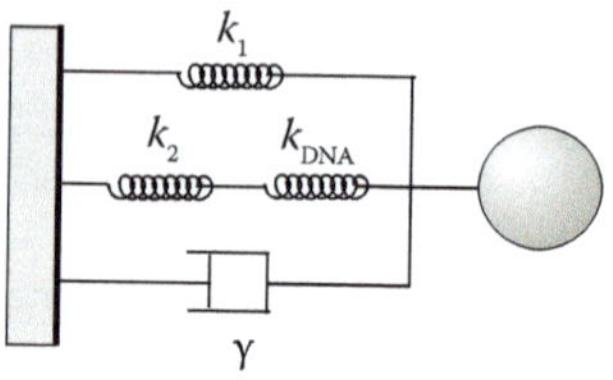

(b)

Figure 5.14 Mechanical elements affecting thermal noise in a two-bead assay. (a) k_1 and k_2 are the elastic constants of the two traps, k_{DNA} comprises the DNA stiffness in series with the stiffness of the linkages connecting the polymer to the beads. γ_1 and γ_2 are the viscous drag coefficients of the two beads. (b) Model of the mechanical elements acting on the left bead in the two-bead assay represented in (a); k_1 and the series of k_{DNA} and k_2 act in parallel on the left microsphere. The viscous drag γ depends on γ_1, γ_2, their hydrodynamic coupling, and the stiffness of the DNA molecule and traps (Moffitt et al. 2006). In case we neglect hydrodynamic coupling between the two beads, γ goes from γ_1 when $k_{DNA} \ll k_{1,2}$ (floppy dumbbell, the movement of the two beads is uncoupled) to $\gamma = \gamma_1 + \gamma_2$ for $k_{DNA} \gg k_{1,2}$ (rigid dumbbell, the movement of the two beads is perfectly coupled).

5.4.2 Instrumental Noise

Thermal fluctuations set a lower limit for high-resolution position measurements, but many sources of instrumental noise deteriorate spatial resolution in optical tweezers. Mechanical oscillations, acoustic noise, thermal expansion, and contraction of the microscope (thermal drifts), as well as laser pointing and intensity instabilities, all negatively affect position measurements. Therefore, high-resolution measurements in surface-coupled geometries (Figures 5.6 and 5.8) require stabilization of both the trapping laser and the microscope stage to reach spatial resolutions near the thermal noise limit. Both active and passive strategies have been adopted in the last decade to address this issue.

Active stabilization of the microscope stage is usually obtained through feedback systems that monitor movements of the stage using a fiducial mark attached to the coverslip surface and compensate such movements using piezo translators. Simple video microscopy can be used, as illustrated in Box 5.3, to monitor movements of a bead stuck to a microscope coverslip and compensate thermal drifts to less than 1 nm in three dimensions (Capitanio et al. 2005). On the contrary, nanometer stability of the trapping laser is achievable with a simple design to minimize air turbulence and laser intensity fluctuations, as described in Section 5.6.5 (Capitanio et al. 2005, 2007b; Monico et al. 2014). Steffen et al. developed such a nanometer-stabilized system to reveal the distribution of binding sites of a single myosin molecule along an actin filament (Steffen et al. 2001).

A much more complex approach is required to reach angstrom stabilization, which is necessary to measure single base pair movements of nucleic acid processing enzymes. Carter et al. used a second laser source and back focal-plane interferometry to monitor the movements of a fiducial mark microfabricated onto the chamber surface and thereby correct thermal drifts using a piezo stage down to ~0.1 nm (Carter et al. 2007). Stabilization of the trapping laser was achieved using an optical fiber to convert laser pointing noise into intensity noise and actively stabilizing it with a feedback loop involving an acousto-optic modulator. Combining stabilization of the sample and the trapping laser, they demonstrated sensitivity to single base-pair steps in a surface-coupled DNA assay (Carter et al. 2007, 2009).

An alternative approach to active stabilization consists of decoupling the biological system under study from the coverslip surface by suspending it between two optical tweezers, formerly stabilized to the required accuracy level. Abbondanzieri et al. reduced laser pointing noise induced by air turbulence by enclosing the laser path within a closed chamber filled with a gas at low refractive index and temperature-stabilizing the experimental apparatus down to 0.1 K variation. They reported ~1 Å position noise when trapping a single bead ($R = 350$ nm) in a stiff trap ($k_{trap} = 1.9$ pN/nm) with a bandwidth $\Delta f = 100$ Hz. In those conditions, Equation 5.13 gives a thermal position noise of ~0.5 Å, close to that reported. By using this ultrastable optical trapping setup, they could probe single base-pair stepping by RNA polymerase (Figure 5.7c) (Abbondanzieri et al. 2005). Using similar double optical tweezers, Cheng et al. could measure single base-pair unwinding of double-stranded RNA by the hepatitis C virus NS3 helicase (Cheng et al. 2011). In this experiment, the setup was temperature stabilized, employed a low-noise laser (Cheng et al. 2010) and the differential detection described in the previous section (Moffitt et al. 2006).

In Section 5.6.5, we investigate the different sources of instrumental noise and give details on strategies that can be adopted to limit them.

5.5 Temporal Resolution

Optical tweezers are affected by various limitations in their capacity to temporally resolve events; the term *temporal resolution* can, thus, refer to different physical quantities and vary depending on the molecule under investigation. Position data can be sampled at high frequencies (hundreds of kilohertz), limited by the bandwidth of the photodetector and electronics. Temporal resolution of the position detector, however, is rarely the limiting factor in optical tweezers measurements.

5.5.1 Relaxation Time

A lower limit on temporal resolution of optical tweezers is given by the relaxation time of the system. In Section 5.2.2.1, we have shown that when a single bead trapped in optical tweezers is perturbed from equilibrium, for example, by protein conformational changes or by trap displacements, it moves exponentially to a new equilibrium position with a time constant (relaxation time) $\tau = \gamma/k$, where γ is the viscous drag coefficient and k the stiffness of the system ($\gamma = \gamma_{bead}$

and $k = k_{trap}$ in that example). Therefore, systems with high stiffness attached to small probes exhibit fast responses to perturbations. If the perturbation develops faster than τ, the bead moves with the same relaxation time τ, filtering out all the movements that occur on shorter time scales.

Let us consider again the arrangement depicted in Figure 5.6a, where a micrometer-sized bead is trapped in optical tweezers and linked to a biological molecule, immersed in a viscous solution. For such a surface-coupled geometry, the stiffness of the surface-coupled molecule and the stiffness of the optical tweezers act in parallel and sum with each other ($k = k_{trap} + k_{motor}$ in Figure 5.12). Therefore, beads attached to rigid molecules display fast temporal responses even in the presence of weak traps. In principle, relaxation times of ~2 μs can be reached with rigid molecules (1 pN/nm) using small beads (200 nm diameter), as shown in Table 5.2. Such values are, however, difficult to reach since, as previously observed in Section 5.4.1, high pretensioning is required to reach stiffness values on the order of 1 pN/nm. Moreover, the force exerted by optical tweezers and the resolution of the position detector both decrease with bead size, making the use of small beads difficult.

Several groups have worked on optimizing the relaxation time of optical tweezers to study biological molecules with high temporal resolution. Nishiyama et al. developed a dark-field position detection scheme with increased signal-to-noise ratio that allowed them to effectively use 200 nm diameter beads in an optical trap. By applying tension to a single kinesin motor proceeding along a microtubule to increase its stiffness, they obtained a response time $\tau < 20$ μs at forces above 3 pN. The high temporal resolution allowed them to detect 4 nm substeps within the 8 nm step of kinesin (Nishiyama et al. 2001). Uemura et al. used the same assay to study the stepping of myosin Va at high spatiotemporal resolution, revealing two pathways for the 36 nm steps, one of them composed by 12 and 24 nm substeps (Uemura et al. 2004). A modified setup was used by Iwaki et al. to tether a single-headed myosin VI to an optically trapped 200 nm bead and rapidly scan the bead along an actin filament (Iwaki et al. 2009). Using this assay, they could observe weak and strong binding of myosin VI heads to actin, and they found that strong binding was greatly enhanced when backward strain was applied.

As observed in Section 5.4.1, trap and molecular stiffness sum up differently in the two-bead assay because of its uncoupling from the coverslip surface. With reference to Figure 5.14, the combined stiffness acting on the left bead is here given by $k = k_1 + \frac{k_2 k_{DNA}}{k_2 + k_{DNA}}$. The stiffness k is always smaller than the combined stiffness of the traps ($k_1 + k_2$) and reaches this limit only for very rigid molecules ($k_{DNA} >> k_2$). Therefore, the relaxation time in this assay is dictated by the stiffness of the two traps, and high trap stiffness helps in both reducing thermal noise and increasing temporal resolution.

5.5.2 Compromise between Spatial and Temporal Resolution—Dead Time

The relaxation time constitutes a lower limit to time resolution that restricts the observable dynamics of biomolecules in optical tweezers. However, thermal fluctuations usually pose a more severe limitation on the time resolution for event detection. In fact, as discussed in Section 5.4.1, to distinguish single base-pair steps with sufficient signal-to-noise ratio, a sufficiently small bandwidth has to be chosen to reduce thermal noise (Equation 5.13). A compromise between spatial and temporal resolution is thus necessary. Therefore, the capacity to detect and measure sub-nm displacements from single molecules requires averaging over several hundreds of milliseconds, thus reducing measurement bandwidth and temporal resolution. The limited temporal resolution sets a lower limit for the briefest detectable event (a molecular motor step in the previous example), which is called the *dead time*. Conformational changes of motor proteins occurring faster than the dead time are not detectable. Equation 5.14 gives the SNR for step detection in a single-bead geometry and can be generalized to a two-bead geometry as in Moffitt et al. (2006). Moffit et al. also gave an explicit expression for the inverse relationship between spatial resolution and temporal resolution (dead time) in a two-bead assay (Moffitt et al. 2008).

The above considerations apply when measuring steps of processive molecular motors and conformational changes or folding trajectories of single molecules in a single- or two-bead assay. A similar analysis can be done when investigating proteins that interact weakly with their binding partners in a three-bead assay, such as nonprocessive motors interacting with their cytoskeletal filament or transcription factors interacting with nonspecific DNA sequences during their target search. Such interactions are characterized by short interaction lifetimes: for example, nonprocessive motors bind to their filament for just one ATP cycle, which lasts for a few milliseconds under physiological conditions. The dead time for the detection of these short interactions is strictly related to the detection method used.

Methods for the detection of interactions of nonprocessive motors in the three-bead assay were first developed in studies on muscle myosin (Figure 5.8a). As briefly introduced in Section 5.3.3, the detection strategy was based on the variation of thermal noise upon binding, which, in turn, depends on the variation of system stiffness, as described by Equation 5.13 in the low-frequency region ($f \ll f_c$) (Molloy et al. 1995; Knight et al. 2001; Smith et al. 2001a). When myosin binds to actin, the system stiffness increases, and thermal fluctuations at low frequencies consequently decrease (as represented by the turquoise area under the blue and red power spectra in Figure 5.13). The ratio between thermal noise in the unbound and bound states $\sigma_u/\sigma_b = k_b/k_u$, where the subscripts u and b respectively indicate quantities measured in the unbound and bound states, can be as high as ~25 by using weak traps and by high pretensioning of the actin filament ($k_u \sim k_1+k_2 \sim 0.04$ pN/nm; $k_b \sim k_m \sim 1$pN/nm) (see Figure 5.15) (Smith et al. 2001a; Capitanio et al. 2006).

In those experiments, the variance of thermal noise is calculated from position data using a time-window consisting of W data points. The signal used to separate bound from unbound events is given by the distance between the variance of the two states $\sigma_u^2 - \sigma_b^2$, while the noise is given by the sum of the standard deviations of those variance states:

$$\mathrm{SNR} = \frac{\sigma_u^2 - \sigma_b^2}{\sigma(\sigma_u^2) + \sigma(\sigma_b^2)} \sim \sqrt{\frac{W}{2}} \frac{\sigma_u^2 - \sigma_b^2}{\sigma_u^2 + \sigma_b^2} \sim \sqrt{\frac{W}{2}} \frac{k_b^2 - k_u^2}{k_b^2 + k_u^2} \tag{5.15}$$

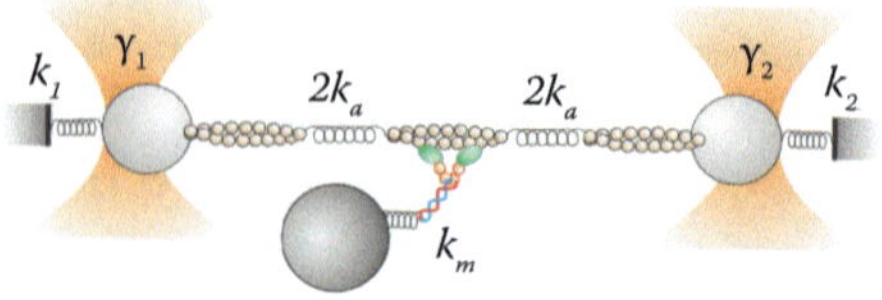

(a)

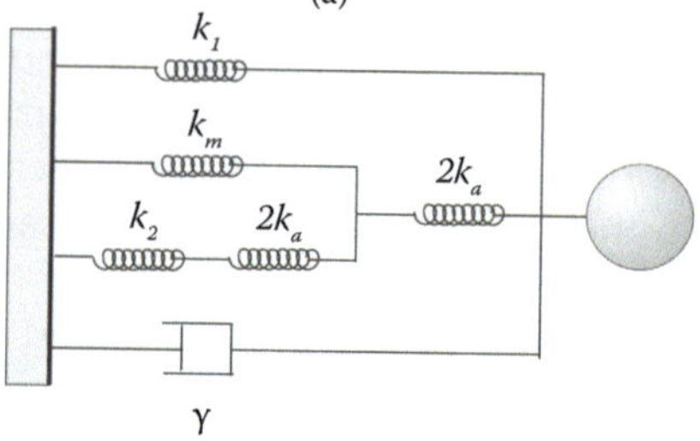

(b)

Figure 5.15 Mechanical elements in a three-bead assay. (a) k_1 and k_2 are the elastic constants of the two traps, $2k_a$ is the stiffness of half actin filament in series with the stiffness of one linkage connecting actin to one of the two beads, k_m the stiffness of the myosin molecule, which binds in the middle of the actin filament. γ_1 and γ_2 are the viscous drag coefficients of the two beads. (b) Model of the mechanical elements acting on the left bead in the three-bead assay represented in (a); the left bead is connected to the left trap k_1 and to half of the actin filament $2k_a$, which is connected in series with the parallel of the myosin molecule k_m and the other half of actin $2k_a$ in series with the left microsphere k_2. The viscous drag γ depends on γ_1, γ_2, their hydrodynamic coupling, and the stiffness of all the elastic elements (Smith 1998; Moffitt et al. 2006). In case we neglect hydrodynamic coupling between the two beads, γ goes from γ_1 when $k_a \ll k_{1,2}$ (floppy dumbbell, the movement of the two beads is uncoupled) to $\gamma = \gamma_1 + \gamma_2$ for $k_a \gg k_{1,2}$, k_m (rigid dumbbell, the movement of the two beads is perfectly coupled). When myosin detaches from actin, the model is still valid with $k_m = 0$.

The first approximation in Equation 5.15 is obtained for $W \gg 1$ (Smith et al. 2001a), while the second is for $f \ll fc$. Although sufficiently high SNR can be obtained for $W > 50$, if the acquisition rate of data points is too high, the above approximation is no longer valid. In fact, $\sigma_u^2 - \sigma_b^2$ and the SNR progressively decrease with frequency, from the value given by Equation 5.15 for $f \ll fc$ down to 0 when the position variance is measured at $f \gg f_c$ (as represented by the pink area in Figure 5.13, which is the same for the blue and red power spectra). For this reason, position variance in this kind of experiment must be calculated using time windows $\Delta t > 5$ ms (Smith et al. 2001a), and events with durations below this value are not detectable. Δt is, thus, the dead time for detection of molecular interactions in a conventional three-bead assay.

Different approaches have been recently developed to overcome such limitation. Veigel et al. developed a technique to decrease the dead time in a three-bead geometry to about 1 ms (Veigel et al. 1999). The idea was to artificially increase the position noise (and thus the signal in Equation 5.15) at a high frequency by oscillating one trap at 1 KHz and detecting the amplitude of oscillation at the other bead, which was greatly reduced upon myosin binding. This method enabled them to detect the onset of each binding event within ~1–2 ms.

An additional complication of the three-bead assay is that it is an unloaded configuration, in which the actin filament is in equilibrium, with the two traps applying equal and opposing forces that sum up to zero. Application of loads in this assay requires putting the system out of equilibrium, for example, by moving the traps after detection of actin–myosin attachment. Such a procedure introduces a *time lag* between myosin attachment and application of the force that is again limited by the dead time (Veigel et al. 2003). Veigel et al. were able to apply a range of loads to a single myosin head with a time lag of ~4–5 ms, which allowed them to investigate the effects of load on the kinetics of the second step of smooth muscle myosin (Veigel et al. 2003) and myosin Va (Veigel et al. 2005). These experiments made a great contribution to the current understanding of how load regulates myosin motors function and how it plays a role in coordination of the two motor domains in processive motors. The several milliseconds lag of this technique, however, did not allow the study of load dependence of faster processes, such as the first step in myosin working stroke. Moreover, techniques capable of clamping force on non-processive motors and weak molecular bonds were still lacking.

5.5.3 Ultrafast Force-Clamp Spectroscopy

A different approach, named *ultrafast force-clamp spectroscopy*, allows application of constant loads between a single intermittently interacting biological polymer and a binding protein with a time lag in the application of load that is only limited by the relaxation time of the system, and with a much improved dead time for interaction detection (Capitanio et al. 2012).

A sketch of the operational principle of the method is shown in Figure 5.16a, in which A is the binding protein, and B is the polymer. A net constant force ($F_{tot} = +\Delta F$) is applied to the bead–polymer–bead complex (dumbbell) through two feedback systems that clamp the force on the two beads to two different values ($-F$ on the left bead and $F+\Delta F$ on the right bead). The dumbbell, thus, moves against viscous drag at constant velocity ($v_u = F_{tot}/\gamma$) when molecules A and B are not bound. The net force is alternated in direction, so that the dumbbell oscillates in a triangular wave fashion within a limited spatial interval (Figure 5.16b). When A binds to B, the force F_{tot} is transferred to the surface-coupled molecule A. The system dynamics is analogous to that described in Section 5.2.2.1 and Equation 5.8, in which a constant force is applied to A and the system exponentially reaches an equilibrium position and stops ($v_b = 0$).

The time taken to transfer the force from the viscous solution to molecule A and stop the dumbbell is the relaxation time of the molecular complex, which also represents the time lag for force application. When applied to the interaction between fast skeletal muscle myosin and actin using highly pretensioned actin filaments and 500 nm diameter beads, the time lag can be as short as ~10 μs (see Table 5.2). For lactose repressor interactions with a weakly tensioned DNA

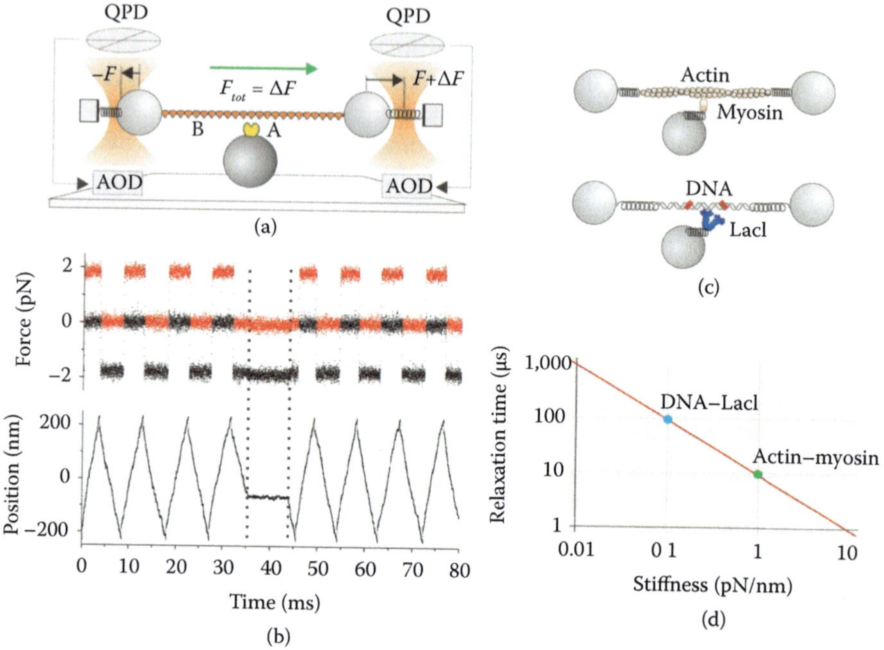

Figure 5.16 Ultrafast force-clamp spectroscopy. (a) The operational principle of the method illustrating constant force $F_{tot} = \Delta F$ applied to the polymer B through two feedback systems clamping the force on the left bead to F and on the right bead to $F + \Delta F$. The force is measured using quadrant detector photodiodes and kept constant by moving the traps via AODs. (b) *Top*: ΔF on left (black) and right (red) beads. *Bottom*: Position of one of the two traps. The force is switched between $+\Delta F$ and $-\Delta F$ to keep the dumbbell within a confined spatial interval (±200 nm). The dumbbell stops when A binds to B (dotted lines). (c) Mechanical model for actin–myosin and DNA-LacI interaction. (d) Relaxation times versus stiffness calculated from the models in panel (c) using 500 nm diameter beads.

(~3 pN), which is characterized by a ~10 times smaller stiffness, the lag is ~100 μs (Figure 5.16c and d). These time lags are very short compared with the duration of typical protein interactions, which, thus, occur under a real force-clamp configuration. Any conformational change occurring after the formation of the molecular bond is also performed under the same constant load (Figure 5.17). Using this system, we could directly apply constant loads before the onset of the first step of myosin and directly measure load dependence of the amplitude of myosin working stroke (Capitanio et al. 2012).

Another advantage of this system is the short dead time. Here, interactions are detected from variations in the dumbbell velocity, which occur with the same short relaxation time of the system, rather than through variations in position noise, which have to be evaluated on a time scale much longer than the relaxation time to give a detectable signal, as discussed in the previous paragraph.

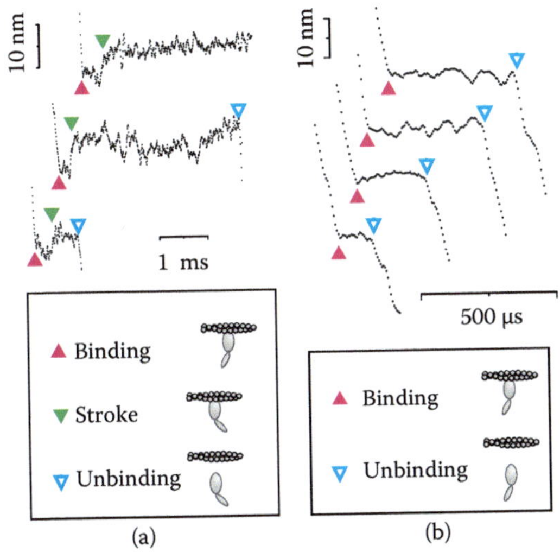

Figure 5.17 Typical interactions in ultrafast force-clamp spectroscopy. (a) Actin-myosin interactions longer than 1 ms showed that the myosin working stroke is developed 0.2–1 ms after attachment. (b) Submillisecond single actin–myosin interactions detected with ultrafast force-clamp spectroscopy. (Filled arrowheads pointing up: Actin–myosin binding. Filled arrowheads pointing down: Myosin working stroke. Open arrowheads pointing down: Actin–myosin detachment).

Therefore, the variation of velocity upon binding ($v_u - v_b = F_{tot}/\gamma$) displays a high signal-to-noise ratio on time scales close to the relaxation time, especially for small beads and high forces. High values of molecular stiffness also increase SNR by reducing thermal noise. For example, the dead time for a rigid actin–myosin dumbbell is ~100 μs at ~5 pN, using 500 nm diameter beads, which allowed the detection of weak actin–myosin interactions (detachment ~5×10^3 s^{-1} at 5 pN), and premature detachment of myosin from actin (~1×10^3 s^{-1} at 5 pN). On Lac repressor interacting with a weakly tensioned DNA, sub–ms interactions are still clearly detectable, allowing the detection of short-lived interactions (dissociating at ~1×10^3 s^{-1} at 4 pN) that are probably involved in Lac repressor facilitated diffusion mechanism (Berg et al. 1981; Monico et al. 2013).

5.5.4 Ensemble Averages

Ensemble averaging is a useful technique when high spatial and temporal resolutions are both required (Veigel et al. 1999; Veigel et al. 2002; Chen et al. 2012). This method enabled distinguishing two steps in the working stroke of myosin Va (Veigel et al. 2002), smooth muscle myosin (Veigel et al. 2003), and skeletal muscle myosin (Capitanio et al. 2006), and excluded the presence of substeps in kinesin (Carter and Cross 2005) (Figure 5.18). Instead of time-averaging position data of single events to reduce thermal noise, as prescribed by Equation 5.13, in ensemble averaging, N interaction events are synchronized at the beginning or at the end of the interaction and averaged point-by-point. Position noise thus scales as $1/\sqrt{N}$

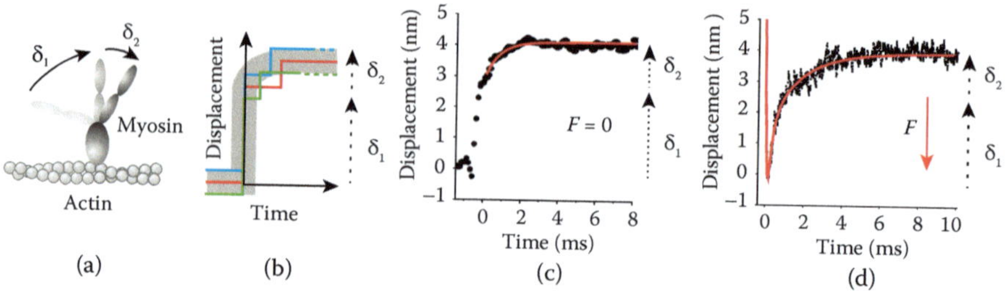

Figure 5.18 Ensemble averages. (a) Myosin working stroke is made of two steps (δ1, δ2). (b) Single actin–myosin interactions (continuous green, red, and blue lines) are composed by two subsequent steps. In ensemble averages, interactions are aligned at their beginning (forward average) or end (backward average, not shown). In the forward average, the last position value of each event is replicated (dotted lines) to match the length of the longest event (red). Ensemble averages (gray lines) are obtained by averaging the events point by point. (c) Ensemble average analysis of interactions between actin and fast skeletal myosin acquired with an unloaded three-bead assay evidenced that a first step (~3.5 nm) occurred very rapidly after myosin binding and was followed by a smaller step (~1 nm) in the same direction (Capitanio et al. 2006). The rate of development of the second step is obtained by fitting the rising phase of the second step with an exponential function (red curve). (d) Ensemble average analysis of interactions between actin and fast skeletal myosin acquired with ultrafast force-clamp spectroscopy confirms that a rapid first step (~4 nm) was followed by a smaller step (~1 nm) (Capitanio et al. 2012). Force is 3.2 pN opposing the working stroke for the data reported in the figure. The increased temporal resolution allows fitting both steps with a double exponential function to obtain their rate of development (red curve).

and can reach the angstrom level for $N \sim 1000$ (Capitanio et al. 2006). Temporal resolution for this kind of average is determined by the accuracy in detecting the time of the beginning or end of the interaction (σ_A). Such accuracy has been evaluated through simulated data (Carter and Cross 2005; Capitanio et al. 2006) or theoretical considerations (Capitanio et al. 2012). For the interaction between fast skeletal muscle myosin and actin, $\sigma_A \sim 300$ μs can be obtained using an unloaded three-bead assay (Capitanio et al. 2006) and in the range 10–50 μs with ultrafast force-clamp spectroscopy (Capitanio et al. 2012). Methods have been developed to interpret the kinetics after an ensemble average and extract the rate constants of the different reactions within a biochemical cycle (Chen et al. 2012).

5.6 Optical Tweezers Setup

Optical tweezers are commonly introduced in a commercial microscope adapted to allow superimposition of the trapping laser with the microscope imaging path. However, optical tweezers can be also built together with the optical microscope as a custom setup. Both approaches have their own advantages and drawbacks.

Commercial microscopes are user friendly and easily allow interfacing with commercial software for image acquisition and device control. Lenses and optics are usually not optimized for their use with near-infrared light as commonly used in optical tweezers for biological applications. The mechanical structure sometimes shows insufficient mechanical stability for measuring nanometer or subnanometer protein conformational changes (see Section 5.4.2) and might be difficult to adapt to optical tweezers operation. Custom setups offer maximum freedom in design but cannot provide the same ease of operation of a commercial microscope. A third option is commercial optical tweezers setups. Among the others, you can find the Nanotracker 2 from JPK Instruments (www.jpk.com) and the C-trap and SuperC-trap from Lumicks (www.lumicks.com). For the price that you pay, you get years of development of optimized hardware and software to control the traps and analyze measurements as well as combined confocal or superresolution microscopy.

Whatever the choice, the basic components are always the same. Following, we give an overview of the instrumentation and optics composing an optical tweezers setup, with a particular focus on describing the basic concepts in optical tweezers design and important features for high-resolution single molecule measurements. Further technical details can be found in several reviews describing optical tweezers design (Fallman and Axner 1997; Smith et al. 1999; Neuman and Block 2004; Dienerowitz et al. 2008).

Figure 5.19 shows a sketch of a typical basic optical tweezers setup. The right-end side of the figure shows the optical microscope pathway, with brightfield illumination and imaging on the lower and upper ends, respectively, fluorescence excitation and detection, and the sample (S) between the objective (O) and the condenser (C). Optical tweezers are represented on the left side, with the trapping laser, beam expansion, and steering optics on the upper part, and position detection on the lower side.

5.6.1 Laser Source and Trapping Beam

When optical tweezers are used to manipulate biological samples, the trapping laser is usually chosen in the near-infrared region of the light spectrum. In fact, absorption of laser light and photodamage has been evaluated in previous works, and few wavelengths in the near-infrared region were found to minimize such effects (Neuman et al. 1999). A 1064 nm wavelength is commonly used because of the availability of high-power solid-state Nd:YAG lasers, even though it is about a factor of 2 more destructive than 830 and 970 nm light, where photodamage exhibits minima. Currently, single-mode diode lasers are available at all the favorable wavelengths, but only at relatively low power. We would like to point out, however, that photodamage is particularly important when working with live cells, whereas, for the purpose of single molecule *in vitro* studies, it is a less important concern. Since the laser source should exhibit high power, good pointing stability, low intensity noise, and high-quality TEM00 spatial mode, solid-state Nd:YAG lasers are still the best option for *in vitro* experiments.

Several optics are introduced in the beam path between the laser source and the microscope objective for optimal optical tweezers operation. With reference to Figure 5.19, after collimation by telescope T1, the laser beam typically

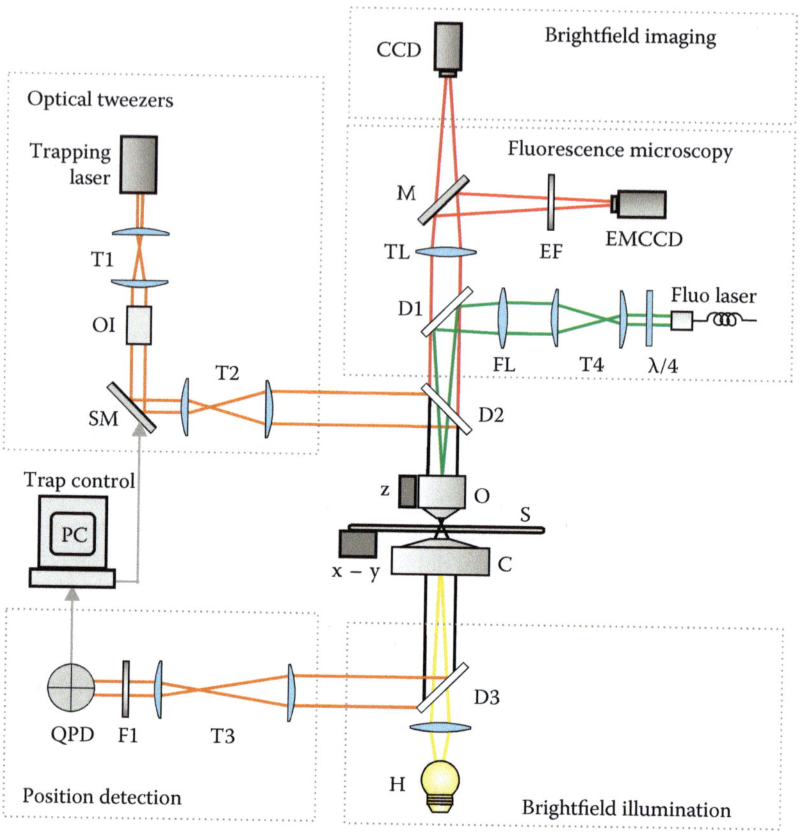

Figure 5.19 An optical tweezers setup. Brightfield illumination for the optical microscope (yellow line) is provided by a halogen lamp (H) and focused in the back focal plane of the condenser (C). The sample (S) can be translated and focused with piezo translators (*x*-*y* and *z*). The objective (O) and the tube lens (TL) project the image of the sample (black and red lines) on a CCD camera (CCD). Optical tweezers (orange line) are inserted and extracted from the optical axis of the microscope through dichroic mirrors (D2 and D3). The trapping beam is collimated by the telescope T1, passes through an optical isolator (OI), and is steered by a steering mirror (SM) controlled by a computer. The beam is expanded by the telescope T2 to match the dimension of the objective pupil. After the condenser, the dimension of the trapping beam is adapted to the quadrant photodiode detector (QPD) through a telescope T3 and filtered by an interference filter F1 to remove unwanted wavelengths. Signals from the QPD are acquired by a computer. Fluorescence excitation is provided by a laser (green line), circularly polarized by a λ/4 waveplate, expanded by the telescope T4, and focused on the back aperture of the objective through a focusing lens (FL). A dichroic mirror D1separates excitation and emission wavelengths, and a movable mirror (M) allows switching between brightfield and fluorescence imaging, which is provided by an electron multiplied camera (EM CCD) after passing through the emission filter (EF).

passes through an optical isolator (OI). Such a device is important to avoid back-reflections of the beam inside the laser cavity, which otherwise cause large random amplitude fluctuations in the laser source.

Since trapping forces originate from the interaction of the dielectric particle and the gradient of light intensity, higher gradient results in higher trapping efficiency. Therefore, the beam dimensions must be adapted to slightly overfill the objective aperture. This allows the operator to fully exploit the objective numerical aperture to achieve maximum beam focusing and light intensity gradient. A second telescope is thus introduced in the optical path for this purpose (T2). For the same reason, the laser source must show a high-quality Gaussian TEM00 mode (i.e., a M quality factor close to 1) to get the smallest waist when focused by the microscope objective.

5.6.2 Beam Steering and Sample Movement

An important feature for optical tweezers is the capability of precisely moving the trap within the sample plane. This can be achieved by steering the laser beam by using a motorized mirror, an acousto-optic deflector, or an electro-optic deflector. The steering device must be precisely positioned in the beam path in order to convert beam angular deflections (ϑ) into trap movements in the sample plane (x). Figure 5.20 illustrates how this conversion is achieved.

The steering device is positioned in the back focal plane (BFP) of the first lens of the telescope T2 and the objective BFP is positioned in the forward focal plane (FFP) of the second lens of T2. Therefore, beam steering occurs on a plane conjugated to the objective BFP (blue dotted lines in Figure 5.20 represent conjugated planes), and rotation of the beam in the objective BFP is converted into trap displacement (x) in the sample plane by the objective lens. Since the objective BFP coincides with the objective back aperture, the beam remains centered on the aperture, independent of ϑ. Thus, proper optics alignment is important to avoid decentering and cutting part of the trapping beam at the objective aperture when the beam is rotated by the steering device.

Important features of the steering device are the maximum and minimum angular deflections (ϑ_{max} and ϑ_{min}) and angular stability (σ_ϑ), which, respectively, determine the maximum and minimum trap displacement (x_{max} and x_{min}) and its pointing stability (σ_x). Using geometrical optics approximation, we obtain the trap displacement at the sample plane (x) as a function of ϑ, the focal length of the objective (fo), and the focal lengths of the telescope (f_1, f_2):

$$x \sim f_O \frac{f_1}{f_2} \vartheta = \frac{f_O}{M} \vartheta, \tag{5.16}$$

where $M = f_2/f_1$ is the magnification of telescope T2. Equation 5.16 relates x_{max}, x_{min}, and σ_x to ϑ_{max}, ϑ_{min}, and σ_ϑ, respectively, and guides us in the choice of a proper steering device given our experimental needs in terms of displacement and stability of the trap in the sample plane. Equation 5.16 also shows that large telescope magnification after the steering device (M) reduces the maximum displacement of the trap in the sample plane; at the same time, it allows smaller

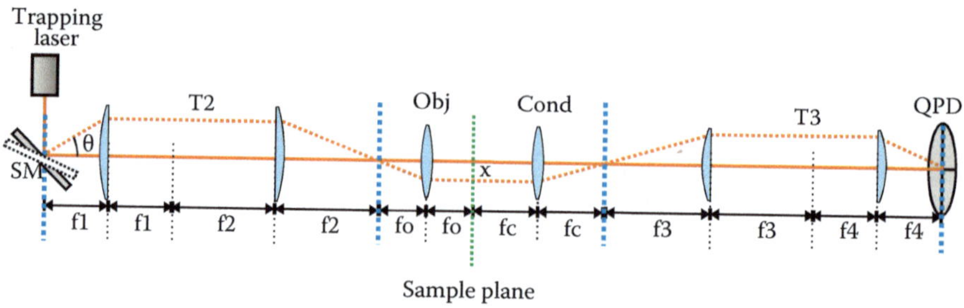

Figure 5.20 Optical path of the trapping beam. Beam steering propagates along the trapping laser path. Dotted blue lines indicate conjugated planes. When the beam direction is rotated by an angle θ by the steering mirror, the beam rotates at the back focal plane of the objective, which converts the rotation into beam translation x at the sample plane. Beam rotation also occurs at the condenser back focal plane and on the QPD.

minimum trap movements and reduces pointing noise. Therefore, the choice of proper magnification of the two telescopes T1 and T2 must take into account, on one hand, the required beam dimension at the objective back aperture and, on the other hand, experimental needs in terms of max/min trap displacements and pointing stability.

Another fundamental feature of the steering device is its deflection speed. Our needs in terms of speed depend again on the experimental configuration, as described in detail in Section 5.3. Electro-optic deflectors offer maximum speed and pointing stability, but limited angular deflection (Valentine et al. 2008). Acousto-optic devices are more commonly used. When driven with stable direct digital synthesizers (DDS) they offer angstrom movements and stability, large trap movements (~50 μm), and fast steering (few μs). Piezo-mirrors or mirrors driven by stepping motors offer larger deflections but with a much slower time response on the millisecond time scale, dictated by the inertia of mechanical movements.

5.6.3 Position Detection

Several position detection methods have been developed during the last decades (Neuman and Block 2004). Here we describe position detection with a quadrant photodiode (QPD) in the back focal plane of the condenser. This system is largely used owing to the high temporal and spatial resolution and capability of 2D and 3D position detection.

In this detection technique, the same laser source is usually used for both trapping and position detection. The trapping laser light, after being focused by the objective in the sample plane, is collected by the condenser and projected onto a QPD, which is positioned in a plane conjugated to the condenser BFP (Figure 5.20). The telescope T3 serves for both creating the conjugated plane and adapting the laser beam to the QPD size. In this configuration, the QPD images the far-field interference between light scattered by the trapped particle and unscattered light (Allersma et al. 1998; Gittes and Schmidt 1998c). The intensity distribution of the interference pattern depends on the position of the bead relative to the trap and the intensity center-of-mass can be measured from differential voltage signals V_x and V_y obtained from the four quadrants currents i_A, i_B, i_C, and i_D, and an operator circuit with gain g (Figure 5.21):

$$V_x = g\frac{(i_B + i_D) - (i_A + i_C)}{i_A + i_B + i_C + i_D}; \quad V_y = g\frac{(i_A + i_B) - (i_C + i_D)}{i_A + i_B + i_C + i_D}; \tag{5.17}$$

These voltage signals are linearly dependent on the bead displacement in the usual range of optical tweezers operation, as discussed in Boxes 5.1 and 5.2. Moreover, the axial position of the bead can be measured from the total light intensity in the back focal plane of the condenser. Also the axial position signal derives from the interference between the light scattered by the trapped particle and unscattered light, as shown by Pralle et al. (Pralle et al. 1999).

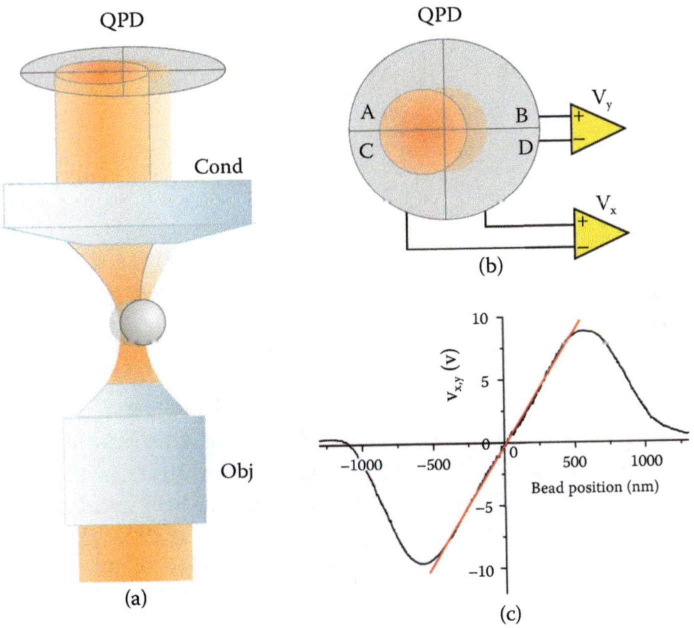

Figure 5.21 Position detection with a QPD positioned in the back focal plane of the condenser. (a) When the bead is located in the center of the trap, the beam is projected in the center of the QPD. As the bead is displaced laterally, the intensity pattern on the QPD moves proportionally. (b) Position signals V_x and V_y are obtained from differential amplifiers. (c) Position signals are proportional to the bead displacement in a range of about ±500 nm from the trap center when using 1 micron diameter beads.

Box 5.2 Multiple Optical Tweezers

Multiple optical tweezers are very useful to manipulate multiple molecules simultaneously, stretch biological polymers, or apply tension on living cells. We previously discussed how double optical tweezers are used in a two- or three-bead configuration to investigate DNA elasticity, DNA processing enzymes, or molecular motors interacting with their cytoskeletal filament. Multiple optical tweezers (MOTs) can be obtained using different methods, but all of them are produced using either time-shared (TS) or continuous (CW) laser beams. The first class of MOT is obtained by rapidly moving a single optical trap between different positions in the sample plane; if the time taken to scan the different trap positions is much smaller than the diffusion time of the trapped particles, the laser beam works as stable multiple optical tweezers (Figure 5.22b). Faster scans imply that more traps can be generated simultaneously and/or that more efficient trapping can be achieved. TS traps

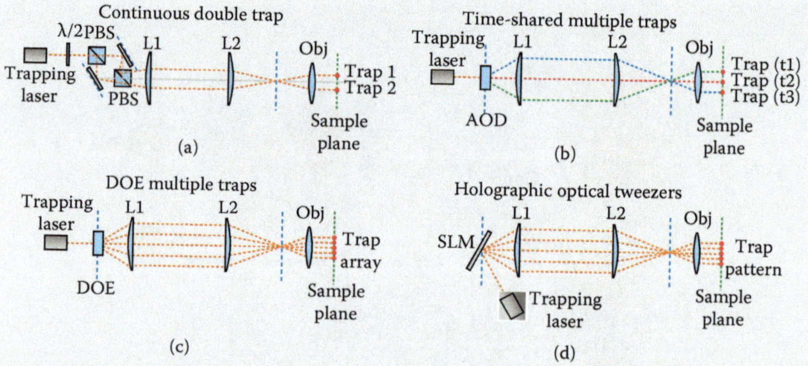

Figure 5.22 Configurations for multiple optical tweezers. (a) Continuous double trap. A single laser beam is split by a polarizing beam-splitter cube (PBS). A λ/2 waveplate allows regulating the laser power in the two arms. Two mirrors, in planes conjugated to the objective back focal plane, rotate the beams, which are then overlapped by a second PBS. The rotation is converted by the objective lens into trap movements in the sample plane. (b) Time-shared multiple traps. An acousto-optic deflector (AOD) rapidly steers the laser beam in multiple angles, which are converted by the objective lens into multiple trap positions in the sample plane. (c) Diffractive optic multiple traps. A diffractive optic placed in a plane conjugated to the objective back focal plane produce an array of laser beam at different angles, which correspond to an array of spots in the sample plane. (d) Holographic optical tweezers. A spatial light modulator (SLM) placed in a plane conjugated to the objective back focal plane produces a pattern of laser beam at different angles, which corresponds to an array of spots in the sample plane.

have been obtained by deflecting the laser beam using piezoelectric mirrors (Mio et al. 2000), galvano mirrors (Sasaki et al. 1991), or acousto-optic deflectors (AODs) (Visscher et al. 1996; Nambiar and Meiners 2002; Guilford et al. 2004; Noom et al. 2007). Depending on the technique used, scanning rates can reach, respectively, 1–2, 10–50, and 10–200 kHz. The generation of the traps is usually controlled from a computer, so that the number of traps, their position, and stiffness can all be modified in real time. TS tweezers are easily built and aligned, since only one laser beam is needed.

CW MOT are obtained by simply dividing a beam into two or more optical paths and then recombining the beams before the objective (Finer et al. 1994); alternatively, two or more laser sources can be combined together (Visscher et al. 1996). This approach is simple when only two traps are needed (Figure 5.22a), but becomes more complicated when multiple traps are required. In those cases, diffractive optical elements or computer-generated holograms can be used to obtain static trap arrays (Figure 5.22c) (Dufresne and Grier 1998; Dufresne et al. 2001), or spatial light modulators that allow real-time control of trap positions in the sample volume (Figure 5.22d) (Liesener et al. 2000; Curtis et al. 2002; Grier 2003).

The double optical tweezers used in the two- and three-bead assays described in Section 5.3 are usually realized by splitting a single laser beam with polarizing beam splitters, resulting in two continuous laser beams (Figure 5.22a). An alternative approach consists in time-sharing the laser beam between the two traps by rapidly moving it between the two positions using AODs (Guilford et al. 2004). Recently, such an approach has been used to build a high-resolution optical trap combined with single-fluorophore sensitivity (see Section 5.7.1) (Comstock et al. 2011). A drawback of the time-shared double trap is that the dumbbell oscillates because of the alternate presence of just one trap that pulls toward it (Capitanio et al. 2007a). The amplitude of the oscillation depends on the relaxation time of the dumbbell in the traps ($\tau \sim 2\gamma_{bead}/2k_{trap}$ for a rigid dumbbell), on the commutation time of the traps, and on the tension imposed to the filament, and usually ranges from ~1 nm at 100 kHz commutation time (close to the limit of AODs) to ~10–20 nm at 10 kHz commutation time. Therefore, the position detection strategy must take into account such oscillations to attain the few-Armstrongs accuracy as reported by Comstock et al.

The signal-to-noise ratio in lateral position detection is maximized when the numerical aperture of the detector is maximized, contrary to axial position detection, which is inversely proportional to the numerical aperture.

5.6.4 Combining Fluorescence Microscopy

The combination of optical tweezers with fluorescence microscopy is very useful whenever a need to visualize the molecules under study or to acquire simultaneous fluorescence and mechanical data emerges. Different illumination schemes for fluorescence microscopy have been adopted depending on the biological application, as described in Section 5.7.2. Here, we briefly illustrate simple wide-field illumination, as schematically depicted in Figure 5.19. Illumination for fluorescence microscopy is usually supplied by one or more laser sources, which provide the high power that is necessary to get detectable signals from single chromophores. Since the laser light is usually linearly polarized, the light polarization must be circularized by a $\lambda/4$ waveplate to maximize excitation of single chromophores, independent of their orientation in the sample plane. The beam is then magnified using a telescope (T4) and focused by a lens (FL) in the back focal plane of the objective to get a collimated beam on the sample plane with uniform intensity within the field of view. A long-pass dichroic mirror (D1) reflects the fluorescence excitation laser and transmits the longer wavelengths emitted by the chromophore. The emitted light is filtered (EF) to efficiently select the emission spectrum and projected onto an EM CCD camera by the tube lens (TL). A motorized mirror (M) allows switching between brightfield and fluorescence microscopy.

The choice of well-separated wavelengths for trapping and fluorescence excitation is necessary for the efficient separation of the two laser beams through dichroic mirrors and filters. The use of a high quantum efficiency electron-multiplied CCD or sCMOS camera is essential to reach the high signal-to-noise ratios necessary for single chromophore detection.

5.6.5 Noise Isolation

Optical tweezers have the capability to measure subnanometer conformational changes of protein enzymes, but, as previously highlighted in Section 5.4.2, many sources of instrumental noise can deteriorate spatial resolution in optical tweezers. Electronic noise in the position detector can limit spatial resolution, but recent photodetectors and electronics allow reaching angstrom-level noise with several KHz bandwidth. Other sources of noise in position measurement are much more prominent and originate from mechanical oscillations, from thermal expansions and contractions in the microscope (thermal drifts), and from laser pointing and intensity instabilities. It is therefore of fundamental importance to isolate the experimental setup from any source of mechanical perturbation and, when this is not fully accomplishable, compensate those perturbations with active feedback systems. Several strategies have been developed in the last decade, using active or passive stabilization strategies, to limit instrumental noise and reach a spatial resolution that approaches the thermal noise limit.

Acoustic noise usually affects frequencies ranging from about 10 Hz to several KHz, while mechanical noise and thermal drifts usually affect lower frequencies. Acoustic noise may also be enhanced by resonances of the microscope mechanical structure or optics supports. The experimental setup is usually mounted on an optical table equipped with active isolators to limit mechanical vibrations. The microscope structure can be mounted over elastomers that absorb mechanical oscillations in the acoustic range of frequencies (Capitanio et al. 2005). Moreover, devices equipped with cooling fans should be put outside of the optical table and the experimental setup should be placed in a room without direct air flow coming from air conditioning or other sources.

The laser trap is also affected by different sources of noise. Air turbulence deteriorates laser pointing stability and can be minimized by enclosing the laser beam path (Capitanio et al. 2005). Air in the laser path enclosure can be replaced with a low refractive index medium to further reduce turbulence (Abbondanzieri et al. 2005). Optical feedback, which generates light intensity fluctuations, can be limited by placing an optical isolator in the laser path near the laser source, as shown in Figure 5.19, or by passing the laser through an optical fiber and/or using active feedback systems (Carter et al. 2007).

Thermal drifts occur on a slow timescale and are quite difficult to reduce below the nanometer level by passive methods (i.e., temperature stabilization), which is really effective only when the whole room is stabilized at the sub-Kelvin level (Abbondanzieri et al. 2005). Otherwise, active feedback systems that compensate thermal drifts by moving the sample stage through piezoelectric translators are more beneficial and simple to implement (Capitanio et al. 2005; Carter et al. 2007). Box 5.3 describes a simple but effective feedback system that uses a video camera and a fiducial mark attached to the microscope coverslip to attain nanometer stabilization of the sample.

5.7 Optical Tweezers Impact in Biology

The last 20 years have seen astonishing progress in force measurements on biological systems, generated new paradigms on single protein function and opened new possibilities to investigate the complex mechanochemical regulation of cellular and molecular complexes. Optical tweezers have reached a spatial resolution of a few angstroms and temporal resolution of a few microseconds, which allowed elucidating details of the molecular mechanisms of molecular motors, DNA processing enzymes, RNA, and protein folding. Complex multiprotein interactions can now be investigated using a combination of single molecule manipulation and imaging tools. Precise pN sensitivity force measurements are now possible in living cells, opening the way to the study of a myriad of biological processes that are directly regulated by force or connected to the mechanical conditions of the cell and its surrounding environment.

5.7.1 High-Resolution Studies of Single Molecular Motors

Improvements in spatial resolution allowed researchers to probe the single steps of enzymatic reactions fundamental for life: transcription by RNA polymerase (RNAP) and translation by the ribosome. As reported in the previous sections, Abbondanzieri et al. developed an ultrastable optical trapping system with angstrom resolution, which they used to monitor transcriptional elongation by single molecules of *Escherichia coli* RNAP (Abbondanzieri et al. 2005) (Figure 5.7c). They showed that RNAP advances along DNA by discrete steps of 3.7 ± 0.6 Å, a distance consistent with the crystallographic spacing between neighboring base pairs in B-DNA. They also determined the force–velocity relationship for transcription and fits to these data were consistent with a Brownian ratchet model.

Box 5.3 Active Nanometer Stabilization Using a Video Camera

Eliminating thermal drifts and low-frequency noise in a microscope and optical tweezers setup is one of the most challenging tasks. A simple and effective way to get nanometer stability on long timescales is by using a bead stuck onto a coverslip surface as a fiducial mark to monitor drifts of the sample, and compensate them using a high-precision piezo stage capable of nanometer or subnanometer movements (Capitanio et al. 2005; Monico et al. 2014). The sample cell can be prepared by spreading micron-sized silica beads dissolved in a nitrocellulose solution onto a microscope coverslip and then attaching the coverslip to a microscope slide using double-stick tape. A silica bead should then be visualized with the focal plane of the objective positioned slightly above the bead and at large magnification to accurately sample its image with a CCD or CMOS camera. The bead image appears as represented in Figure 5.23 after image inversion and threshold

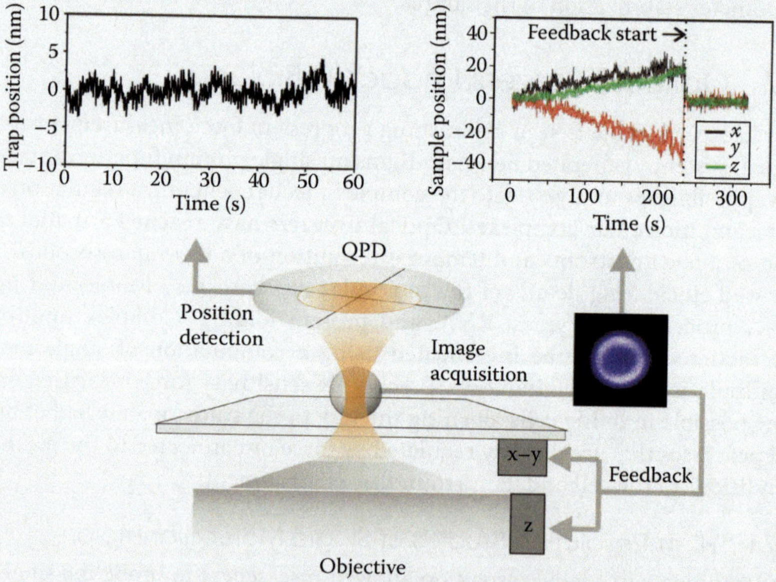

Figure 5.23 Nanometer stabilization of optical tweezers. A bead stuck onto the coverslip surface is used as a reference to monitor the sample position (top right graph), which normally drifts by several nanometers in a few minutes. A feedback system driving piezo translators stabilizes the sample to better than 1 nm. Movements of the trap relative to the sample can be monitored with the QPD (top left graph), when the trap is overlapped on the bead stuck onto the coverslip.

filtering to cut off the background noise outside the bead. The bead x and y coordinates can be obtained from the image centroid, while the z position can be obtained from the ratio between the intensity of the pixels inside the bead diffraction ring and those located outside. A feedback loop that acquires images of the bead, calculates x, y, and z coordinates, and sends a proportional correction signal to the piezo stage can thus compensate sample drifts. After proper calibration of the system and regulation of the feedback gain, the optical microscope can be stabilized to better than 1 nm using this approach (Capitanio et al. 2005) (see Figure 5.23). The range of noise frequencies on which the feedback system operates depends on the acquisition rate of the camera; for thermal drifts and low-frequency noise correction, a slow video-rate camera operating at 25 Hz is sufficient.

When performing optical tweezers experiments on surface-coupled molecules, it is also fundamental to check that the relative position of the feedback-stabilized sample and the optical traps is maintained at the nanometer level. To this end, an experimental procedure is represented in Figure 5.23: the trapping beam is positioned onto a surface-attached bead, which is stabilized by the feedback loop, and the relative movements of the laser and the bead are measured through the QPD. When the prescriptions described in Section 5.6.5 are fulfilled, nanometer stability can be achieved.

This breakthrough opened the way to detailed studies of the molecular mechanisms of RNAP and how gene expression is regulated during transcription. With the use of the same optical-trapping assay, Greenleaf et al. investigated folding of single nascent RNAs containing pbuE adenine riboswitch aptamers (Greenleaf et al. 2008). Riboswitches are important elements of mRNA structure that regulate genes through structural changes in ligand-binding RNA aptamers. In a recent study, Fazal et al. assembled a 32-protein, 1.5-megadalton preinitiation complex of RNA polymerase II, and observed subsequent initiation transcription processes in real time in a double-trap assay (Fazal et al. 2015). Recent studies demonstrate that it is possible to assemble complex macromolecules *in vitro* and investigate their properties with high resolution at a single molecule level.

Another set of extremely challenging experiments demonstrated the possibility of studying translation by the ribosome with optical tweezers (Figure 5.9b). Those experiments revealed that translation occurs through successive translocation-and-pause cycles, with each translocation step measuring three bases (i.e., one codon). This approach allowed investigating the dynamics of ribosome translation and analyzing the time the ribosome spends at each codon, thus revealing that there are three substeps in each step. Pause lengths, and thus the overall rate of translation, depend on the secondary structure of the mRNA, which is destabilized by an applied force, resulting in decrease of pause durations, but not of translocation times. The same group developed an experimental system to investigate the folding of single ribosome-bound stalled nascent polypeptides with optical tweezers (Kaiser et al. 2011). This study directly showed that the ribosome affects protein folding and suggested that the ribosome not only decodes the genetic information and synthesizes polypeptides, but also promotes efficient attainment of the native state.

Improvements in temporal resolution allowed researchers to dissect the steps of chemomechanical transduction in a number of molecular motors, giving new insight into their mechanism of functioning. Myosin is one of the prototypical motor proteins, and it is involved in numerous biological processes such as muscle contraction, intracellular trafficking, cell movements, and signal transduction (Mermall et al. 1998). Skeletal muscle myosin is one of the most long-studied motor proteins, although a detailed analysis of its load dependence has been elusive for a long time to single molecule studies owing to its rapid chemomechanical cycle. Several principles of myosin function were first discovered on slower myosins belonging to different classes and, only recently, generalized to skeletal muscle myosin when new techniques with improved time resolution become available. In 1999, Veigel et al. demonstrated that myosin I, a widely expressed, single-headed, and membrane-associated motor produces its working stroke in two steps (Veigel et al. 1999). The second step of myosin, which was later discovered to be a common feature of the myosin family, regulates myosin kinetics through its load dependence and has emerged to play an important role in the functioning of different myosin (Batters and Veigel 2016). Myosin I was showed to respond to small resisting loads (<2 pN) by dramatically increasing the actin-attachment lifetime more than 75-fold (Laakso et al. 2008). This impressive tension sensitivity supports a role for myosin I as a molecular force sensor adapted to generate and sustain tension for extended time periods, rather than to rapidly transport cargos. On the contrary, myosin V is a double-headed efficient cargo transporter, and its high processivity has been explained by the load dependence of its second step (Veigel et al. 2005). In fact, when myosin V is attached to actin, intramolecular strain between the two myosin heads can significantly enhance the probability of the trail head detaching first, causing a strong increase in the number of forward steps over a system with no strain dependence. A second step was also observed in skeletal muscle myosin thanks to refinements in the temporal resolution of a three-bead assay (Capitanio et al. 2006). In that work, we demonstrated that the kinetics of this second step vary largely between different isoforms of skeletal muscle myosin, accounting for the variation, in different fiber types, of the rate of the cross-bridge cycle, which is tuned to the specific cellular function. Some years later we could finally find a technical solution to apply controlled loads to a single myosin molecule before the onset of its working stroke, thus opening the way to studies of load dependence of the complete myosin working stroke (Capitanio et al. 2012). In that paper, we could directly observe that the amplitude of the working stroke of skeletal muscle myosin is reduced by an opposing force, whereas its duration is slowed down at high ATP concentrations in a physiologic force range (<5 pN). These results support the idea that the contraction velocity of a single myosin molecule decreases with force–accordingly with the force-velocity curve of a whole muscle cell.

5.7.2 Combining Optical Manipulation and Imaging of Single Molecules

A further evolution of optical tweezers techniques consists in the development of a new generation of setups combining several single molecule methodologies to allow, for example, simultaneous single molecule localization using fluorescence microscopy and manipulation with optical tweezers. The first examples of these types of combined technologies consisted of a laser tweezers apparatus used to manipulate a biopolymer (e.g., DNA or actin) combined with an evanescent wave used for excitation and detection of single fluorophores. These setups were used to simultaneously measure mechanical output and ATPase reaction in single myosin molecules (Ishijima et al. 1998), or to image and measure the interactions of RNA polymerase (fluorescently labeled) with a DNA molecule suspended between two optical traps (Harada et al. 1999). These pioneering works combined manipulation of single molecules with detection of single chromophores. The technological developments in video camera sensitivity, allowed some years later the combination of optical tweezers with fluorescence imaging with 1 nm accuracy (FIONA) (Capitanio et al. 2007b). A growing set of tools combining optical tweezers with wide-field fluorescence microscopy techniques for the detection and localization of single molecules have been developed in the last years (Lang et al. 2004; Capitanio et al. 2007b; Monico et al. 2014). These tools have been mainly applied to the study of nucleic acids structure (Hohng et al. 2007; van Mameren et al. 2009) and DNA binding proteins and processing enzymes (Candelli et al. 2011; Zhou et al. 2011; Monico et al. 2014). More recently, different illumination configurations have been developed to selectively excite labeled proteins interacting with a DNA molecule suspended between two optically trapped beads. Comstock et al. developed a time-shared ultrahigh-resolution dual-optical trap interlaced with a confocal fluorescence microscope. The time-shared dual-optical trap was interlaced with a confocal fluorescence microscope, so that optical traps and fluorescence excitation were never both on simultaneously. This solution is useful to avoid enhanced photobleaching due to absorption of the near-infrared trapping light while the chromophore is in the excited state (Dijk et al. 2004). Thanks to this solution, they observed individual single fluorophore-labeled DNA oligonucleotides interacting with complementary DNA and coincident angstrom-scale changes in tether extension (Comstock et al. 2011). Heller et al. combined optical tweezers with stimulated emission depletion (STED) fluorescence microscopy. Proteins on DNA could be imaged with a resolution of 50 nm, a sixfold resolution improvement over that of confocal microscopy, allowing the visualization of individual DNA-binding proteins on densely covered DNA and in the presence of high protein concentrations.

5.7.3 Going Inside the Cell

Besides the numerous and fruitful applications of optical tweezers to the study of single biological molecules *in vitro*, optical tweezers have been applied to live-cell studies as well, and recent developments promise to further extend their usability in this area. Optical tweezers can be used as efficient tools to mechanically stimulate cells or specific membrane proteins (Wang et al. 2005), measure mechanical properties of the cell (Choquet et al. 1997), quantify strength of molecular bonds between membrane proteins and binding partners (Litvinov et al. 2002), or investigate mechanical properties of molecular motors in living cells (Blehm et al. 2013). Several reviews illustrate the application of optical tweezers to live cells (Veigel and Schmidt 2011; Oddershede 2012).

Although manipulating cells from the outer membrane is the most common use, direct manipulation and force measurement inside living cells with optical tweezers have seen great progress recently. One of the main issues for quantitative force measurements in living cells with optical tweezers is the calibration of the trap stiffness k_{trap}. In fact, opposite to *in vitro* experiments or experiments on the cell membrane, in which the viscous properties of the medium are well known, experiments in living cells are complicated by the viscoelastic properties of the cytoplasm, which are *a priori* unknown. Another major problem in this kind of experiment is the variable size of endogenous lipid droplets and vesicles that are usually employed as force probes. Since k_{trap} varies largely with size in the micrometer to submicrometer range (see Table 5.1), it is extremely difficult to precisely calibrate traps for cargoes of unknown size. Barak et al. developed a calibration method to precisely measure the force generated by motor proteins on single cargos of unknown size in cell extracts (Barak et al. 2013). Rai et al. recently demonstrated optical trapping at single molecule resolution inside cells (Rai et al. 2013). They used phagocytosed latex beads of uniform size to circumvent the calibration uncertainty. Fischer et al. developed a calibration method that allows obtaining all the unknown variables for a correct calibration of optical tweezers inside cells (Fischer and Berg-Sorensen 2007; Fischer et al. 2010). Two recent articles exploited and further extended this technique to measure stall force of dynein and kinesin *in vivo*. Hendricks et al. developed a calibration method taking into account any variations among cargoes and local viscoelastic properties of the cytoplasm. They examined latex beads phagocytosed into living mammalian macrophages and found great differences in the characteristics of forces generated by teams of plus- (kinesin) and minus- (dynein) end–directed motors (Hendricks et al. 2012). Blehm et al. used a method that allows *in vivo* calibration of optical tweezers on lipid vesicles and phagocytosed polystyrene beads to study kinesin and dynein stall forces during intracellular transport (Blehm et al. 2013).

References

Abbondanzieri EA, Greenleaf WJ, Shaevitz JW, Landick R, Block SM (2005) Direct observation of base-pair stepping by RNA polymerase. *Nature* 438:460–465.

Allersma MW, Gittes F, deCastro MJ, Stewart RJ, Schmidt CF (1998) Two-dimensional tracking of ncd motility by back focal plane interferometry. *Biophys J* 74:1074–1085.

Arya M, Kolomeisky AB, Romo GM, Cruz MA, Lopez JA, Anvari B (2005) Dynamic force spectroscopy of glycoprotein Ib-IX and von Willebrand factor. *Biophys J* 88:4391–4401.

Ashkin A (1992) Forces of a single-Beam gradient laser trap on a dielectric sphere in the ray optics regime. *Biophys J* 61:569–582.

Ashkin A, Dziedzik JM, Bjorkholm JE, Chu S (1986) Observation of a single-Beam gradient force optical trap for dielectric particles. *Opt Lett* 11:288–290.

Barak P, Rai A, Rai P, Mallik R (2013) Quantitative optical trapping on single organelles in cell extract. *Nat Meth* 10:68–70.

Batchelor GK (1967) *An introduction to fluid dynamics.* Cambridge: U.P.

Batters C, Veigel C (2016) Mechanics and activation of unconventional myosins. *Traffic* 17:860–871.

Berg OG, Winter RB, von Hippel PH (1981) Diffusion-driven mechanisms of protein translocation on nucleic acids. 1. Models and theory. *Biochemistry* 20:6929–6948.

Bianco P, Bongini L, Melli L, Dolfi M, Lombardi V (2011) PicoNewton-millisecond force steps reveal the transition kinetics and mechanism of the double-stranded DNA elongation. *Biophys J* 101:866–874.

Bianco P, Nagy A, Kengyel A, Szatmari D, Martonfalvi Z, Huber T, Kellermayer MS (2007) Interaction forces between F-actin and titin PEVK domain measured with optical tweezers. *Biophys J* 93:2102–2109.

Blehm BH, Schroer TA, Trybus KM, Chemla YR, Selvin PR (2013) In vivo optical trapping indicates kinesin's stall force is reduced by dynein during intracellular transport. *Proc Natl Acad Sci U S A* 110:3381–3386.

Buckley CD, Tan JY, Anderson KL, Hanein D, Volkmann N, Weis WI, Nelson WJ, Dunn AR (2014) The minimal cadherin-catenin complex binds to actin filaments under force. *Science* 346:1254211.

Candelli A, Wuite GJ, Peterman EJ (2011) Combining optical trapping, fluorescence microscopy and micro-Fluidics for single molecule studies of DNA-protein interactions. *Phys Chem Chem Phys: PCCP* 13:7263–7272.

Capitanio M, Canepari M, Cacciafesta P, Lombardi V, Cicchi R, Maffei M, Pavone FS, Bottinelli R (2006) Two independent mechanical events in the interaction cycle of skeletal muscle myosin with actin. *Proc Natl Acad Sci U S A* 103:87–92.

Capitanio M, Canepari M, Maffei M, Beneventi D, Monico C, Vanzi F, Bottinelli R, Pavone FS (2012) Ultrafast force-clamp spectroscopy of single molecules reveals load dependence of myosin working stroke. *Nat Meth* 9:1013–1019.

Capitanio M, Cicchi R, Pavone FS (2005) Position control and optical manipulation for nanotechnology applications. *Eur Phys J B* 46:1–8.

Capitanio M, Cicchi R, Pavone FS (2007a) Continuous and time-shared multiple optical tweezers for the study of single motor proteins. *Optics Lasers Eng* 45:450–457.

Capitanio M, Maggi D, Vanzi F, Pavone F (2007b) Fiona in the trap: The advantages of combining optical tweezers and fluorescence. *J Opt A: Pure Appl Opt* 9:S157.

Capitanio M, Pavone FS (2013) Interrogating biology with force: Single molecule high-resolution measurements with optical tweezers. *Biophys J* 105:1293–1303.

Capitanio M, Romano G, Ballerini R, Giuntini M, Pavone FS, Dunlap D, Finzi L (2002) Calibration of optical tweezers with differential interference contrast signals. *Review Sci Instrum* 73:1687–1696.

Capitanio M, Vanzi F, Broggio C, Cicchi R, Normanno D, Romano G, Sacconi L, Pavone FS (2004) Exploring molecular motors and switches at the single-molecule level. *Micr Res Tech* 65:194–204.

Carter AR, King GM, Ulrich TA, Halsey W, Alchenberger D, Perkins TT (2007) Stabilization of an optical microscope to 0.1 nm in three dimensions. *Appl Opt* 46:421–427.

Carter AR, Seol Y, Perkins TT (2009) Precision surface-Coupled optical-Trapping assay with one-Basepair resolution. *Biophys J* 96:2926–2934.

Carter BC, Vershinin M, Gross SP (2008) A comparison of step-detection methods: How well can you do? *Biophys J* 94:306–319.

Carter NJ, Cross RA (2005) Mechanics of the kinesin step. *Nature* 435:308–312.

Cecconi C, Shank EA, Bustamante C, Marqusee S (2005) Direct observation of the three-state folding of a single protein molecule. *Science* 309: 2057–2060.

Chemla YR, Aathavan K, Michaelis J, Grimes S, Jardine PJ, Anderson DL, Bustamante C (2005) Mechanism of force generation of a viral DNA packaging motor. *Cell* 122:683–692.

Chen C, Greenberg MJ, Laakso JM, Ostap EM, Goldman YE, Shuman H (2012) Kinetic schemes for post-synchronized single molecule dynamics. *Biophys J* 102:L23–L25.

Cheng W, Arunajadai SG, Moffitt JR, Tinoco I, Jr., Bustamante C (2011) Single-base pair unwinding and asynchronous RNA release by the hepatitis C virus NS3 helicase. *Science* 333:1746–1749.

Cheng W, Hou X, Ye F (2010) Use of tapered amplifier diode laser for biological-friendly high-Resolution optical trapping. *Opt Lett* 35:2988–2990.

Choquet D, Felsenfeld DP, Sheetz MP (1997) Extracellular matrix rigidity causes strengthening of integrin-Cytoskeleton linkages. *Cell* 88:39–48.

Comstock MJ, Ha T, Chemla YR (2011) Ultrahigh-resolution optical trap with single-fluorophore sensitivity. *Nat Meth* 8:335–340.

Curtis JE, Koss BA, Grier DG (2002) Dynamic holographic optical tweezers. *Opt Commun* 207:169–175.

Dame RT, Noom MC, Wuite GJL (2006) Bacterial chromatin organization by H-NS protein unravelled using dual DNA manipulation. *Nature* 444:387–390.

de Messieres M, Chang JC, Brawn-Cinani B, La Porta A (2012) Single-Molecule study of G-Quadruplex disruption using dynamic force spectroscopy. *Phys Rev Lett* 109:058101.

deCastro MJ, Fondecave RM, Clarke LA, Schmidt CF, Stewart RJ (2000) Working strokes by single molecules of the kinesin-related microtubule motor ncd. *Nat Cell Biol* 2:724–729.

Dienerowitz M, Mazilu M, Dholakia K (2008) Optical manipulation of nanoparticles: A review. *J Nanophotonics* 2:021875.

Dijk MA, Kapitein LC, Mameren J, Schmidt CF, Peterman EJ (2004) Combining optical trapping and single-molecule fluorescence spectroscopy: Enhanced photobleaching of fluorophores. *J Phys Chem B* 108:6479–6484.

Dudko OK, Hummer G, Szabo A (2008) Theory, analysis, and interpretation of single-molecule force spectroscopy experiments. *Proc Natl Acad Sci U S A* 105:15755–15760.

Dufresne ER, Grier DG (1998) Optical tweezer arrays and optical substrates created with diffractive optics. *Review Sci Instrum* 69:1974–1977.

Dufresne ER, Spalding GC, Dearing MT, Sheets SA, Grier DG (2001) Computer-generated holographic optical tweezer arrays. *Review Sci Instrum* 72:1810–1816.

Dumont S, Cheng W, Serebrov V, Beran RK, Tinoco I, Pyle AM, Bustamante C (2006) RNA translocation and unwinding mechanism of HCVNS3 helicase and its coordination by ATP. *Nature* 439:105–108.

Elms PJ, Chodera JD, Bustamante CJ, Marqusee S (2012) Limitations of constant-force-feedback experiments. *Biophys J* 103:1490–1499.

Evans E (2001) Probing the relation between force - Lifetime - and chemistry in single molecular bonds. *Annu Rev Bioph Biom* 30:105–128.

Fallman E, Axner O (1997) Design for fully steerable dual-trap optical tweezers. *Appl Opt* 36:2107–2113.

Fazal FM, Meng CA, Murakami K, Kornberg RD, Block SM (2015) Real-time observation of the initiation of RNA polymerase II transcription. *Nature* 525:274–277.

Finer JT, Simmons RM, Spudich JA (1994) Single myosin molecule mechanics: Piconewton forces and nanometer steps. *Nature* 368:113–119.

Fischer M, Berg-Sorensen K (2007) Calibration of trapping force and response function of optical tweezers in viscoelastic media. *J Opt a-Pure Appl Op* 9:S239–S250.

Fischer M, Richardson AC, Reihani SN, Oddershede LB, Berg-Sorensen K (2010) Active-passive calibration of optical tweezers in viscoelastic media. *Review Sci Instrum* 81:015103.

Garcia-Garcia C, Frieda KL, Feoktistova K, Fraser CS, Block SM (2015) Factor-dependent processivity in human eIF4A DEAD-box helicase. *Science* 348:1486–1488.

Gebhardt JC, Bornschlogl T, Rief M (2010) Full distance-resolved folding energy landscape of one single protein molecule. *Proc Natl Acad Sci U S A* 107:2013–2018.

Ghislain L, Switz NA, Webb WW (1994) Measurement of small forces using an optical trap. *Rev Sci Instrum* 65:2762–2768.

Gittes F, Schmidt CF (1998a) Thermal noise limitations on micromechanical experiments. *Eur Biophys J Biophy* 27:75–81.

Gittes F, Schmidt CF (1998b) Signals and noise in micromechanical measurements. *Meth Cell Biol* 55:129–156.

Gittes F, Schmidt CF (1998c) Interference model for back-focal-plane displacement detection in optical tweezers. *Optics Lett* 23:7–9.

Greenberg MJ, Lin T, Goldman YE, Shuman H, Ostap EM (2012) Myosin IC generates power over a range of loads via a new tension-sensing mechanism. *Proc Natl Acad Sci U S A* 109:E2433–E2440.

Greenberg MJ, Shuman H, Ostap EM (2014) Inherent force-Dependent properties of beta-Cardiac myosin contribute to the force-Velocity relationship of cardiac muscle. *Biophys J* 107:L41–L44.

Greenleaf WJ, Frieda KL, Foster DA, Woodside MT, Block SM (2008) Direct observation of hierarchical folding in single riboswitch aptamers. *Science* 319:630–633.

Greenleaf WJ, Woodside MT, Block SM (2007) High-resolution, single-molecule measurements of biomolecular motion. *Annu Rev Biophys Biomol Struct* 36:171–190.

Greenleaf WJ, Woodside MT, Abbondanzieri EA, Block SM (2005) Passive all-optical force clamp for high-resolution laser trapping. *Phys Rev Lett* 95:208102.

Grier DG (2003) A revolution in optical manipulation. *Nature* 424:810–816.

Guilford WH, Tournas JA, Dascalu D, Watson DS (2004) Creating multiple time-shared laser traps with simultaneous displacement detection using digital signal processing hardware. *Anal Biochem* 326:153–166.

Guo B, Guilford WH (2006) Mechanics of actomyosin bonds in different nucleotide states are tuned to muscle contraction. *Proc Natl Acad Sci U S A* 103:9844–9849.

Hansen PM, Tolic-Norrelykke IM, Flyvbjerg H, Berg-Sorensen K (2006) Tweezercalib 2.1: Faster version of MatLab package for precise calibration of optical tweezers. *Comput Phys Commun* 175:572–573.

Harada Y, Funatsu T, Murakami K, Nonoyama Y, Ishihama A, Yanagida T (1999) Single-Molecule imaging of RNA polymerase-DNA interactions in real time. *Biophys J* 76:709–715.

Hecht E (2002) *Optics*. 4th Edition. Reading, MA: Addison-Wesley.

Hendricks AG, Holzbaur EL, Goldman YE (2012) Force measurements on cargoes in living cells reveal collective dynamics of microtubule motors. *Proc Natl Acad Sci U S A* 109:18447–18452.

Hohng S, Zhou R, Nahas MK, Yu J, Schulten K, Lilley DM, Ha T (2007) Fluorescence-force spectroscopy maps two-dimensional reaction landscape of the holliday junction. *Science* 318:279–283.

Howard J (2001) Mechanics of motor proteins and the cytoskeleton. MA: Sinauer Associates, Inc. Publisher, Sunderland, MA.

Hulst HCvd (1981) *Light scattering by small particles*. New York, NY: Dover Publications.

Ishijima A, Kojima H, Funatsu T, Tokunaga M, Higuchi H, Tanaka H, Yanagida T (1998) Simultaneous observation of individual ATPase and mechanical events by a single myosin molecule during interaction with actin. *Cell* 92:161–171.

Iwaki M, Iwane AH, Shimokawa T, Cooke R, Yanagida T (2009) Brownian search-and-catch mechanism for myosin-VI steps. *Nat Chem Biol* 5:403–405.

Jaalouk DE, Lammerding J (2009) Mechanotransduction gone awry. *Nat Rev Mol Cell Biol* 10:63–73.

Kaiser CM, Goldman DH, Chodera JD, Tinoco I, Bustamante C (2011) The ribosome modulates nascent protein folding. *Science* 334:1723–1727.

Kim J, Zhang CZ, Zhang X, Springer TA (2010) A mechanically stabilized receptor-ligand flex-bond important in the vasculature. *Nature* 466: 992–995.

Knight AE, Veigel C, Chambers C, Molloy JE (2001) Analysis of single-molecule mechanical recordings: Application to acto-myosin interactions. *Prog Biophys Mol Biol* 77:45–72.

Laakso JM, Lewis JH, Shuman H, Ostap EM (2008) Myosin I can act as a molecular force sensor. *Science* 321:133–136.

Laakso JM, Lewis JH, Shuman H, Ostap EM (2010) Control of myosin-I force sensing by alternative splicing. *Proc Natl Acad Sci U S A* 107:698–702.

Landau LD, Lifshit͡s EM, Pitaevskiĭ LP (1980) *Statistical physics*, 3d rev. and enl. Edition. Oxford; New York: Pergamon Press.

Lang MJ, Asbury CL, Shaevitz JW, Block SM (2002) An automated two-dimensional optical force clamp for single molecule studies. *Biophys J* 83:491–501.

Lang MJ, Fordyce PM, Engh AM, Neuman KC, Block SM (2004) Simultaneous, coincident optical trapping and single-molecule fluorescence. *Nat Meth* 1:133–139.

Lewalle A, Steffen W, Stevenson O, Ouyang Z, Sleep J (2008) Single-Molecule measurement of the stiffness of the rigor myosin head. *Biophys J* 94:2160–2169.

Liesener J, Reicherter M, Haist T, Tiziani HJ (2000) Multi-functional optical tweezers using computer-generated holograms. *Opt Commun* 185:77–82.

Liphardt J, Onoa B, Smith SB, Tinoco IJ, Bustamante C (2001) Reversible unfolding of single RNA molecules by mechanical force. *Science* 292:733–737.

Litvinov RI, Shuman H, Bennett JS, Weisel JW (2002) Binding strength and activation state of single fibrinogen-integrin pairs on living cells. *Proc Natl Acad Sci U S A* 99:7426–7431.

Marshall BT, Long M, Piper JW, Yago T, McEver RP, Zhu C (2003) Direct observation of catch bonds involving cell-adhesion molecules. *Nature* 423:190–193.

Mermall V, Post PL, Mooseker MS (1998) Unconventional myosins in cell movement, membrane traffic, and signal transduction. *Science* 279:527–533.

Mio C, Gong T, Terray A, Marr WM (2000) Design of a scanning laser optical trap for multiparticle manipulation. *Rev Sci Instrum* 71:2196–2200.

Moffitt JR, Chemla YR, Izhaky D, Bustamante C (2006) Differential detection of dual traps improves the spatial resolution of optical tweezers. *Proc Natl Acad Sci U S A* 103:9006–9011.

Moffitt JR, Chemla YR, Smith SB, Bustamante C (2008) Recent advances in optical tweezers. *Ann Rev Biochem* 77:205–228.

Molloy JE, Burns JE, Kendrick-Jones J, Tregear RT, White DCS (1995) Movement and force produced by a single myosin head. *Nature* 378:209–212.

Monico C, Belcastro G, Vanzi F, Pavone FS, Capitanio M (2014) Combining single-molecule manipulation and imaging for the study of protein–DNA interactions. *J Vis Exp: JoVE* 90:51446.

Monico C, Capitanio M, Belcastro G, Vanzi F, Pavone FS (2013) Optical methods to study protein–DNA interactions in vitro and in living cells at the single-Molecule level. *Int J Mol Sci* 14:3961–3992.

Nambiar R, Gajraj A, Meiners JC (2004) All-optical constant-force laser tweezers. *Biophys J* 87:1972–1980.

Nambiar R, Meiners JC (2002) Fast position measurements with scanning line optical tweezers. *Opt Lett* 27:836–838.

Neuman KC, Block SM (2004) Optical trapping. *Rev Sci Instrum* 75:2787–2809.

Neuman KC, Chadd EH, Liou GF, Bergman K, Block SM (1999) Characterization of photodamage to Escherichia coli in optical traps. *Biophys J* 77:2856–2863.

Neuman KC, Nagy A (2008) Single-molecule force spectroscopy: optical tweezers, magnetic tweezers and atomic force microscopy. *Nat Meth* 5:491–505.

Nishiyama M, Muto E, Inoue Y, Yanagida T, Higuchi H (2001) Substeps within the 8-nm step of the ATPase cycle of single kinesin molecules. *Nat Cell Biol* 3:425–428.

Noom MC, van den Broek B, van Mameren J, Wuite GJ (2007) Visualizing single DNA-bound proteins using DNA as a scanning probe. *Nat Meth* 4:1031–1036.

Oddershede LB (2012) Force probing of individual molecules inside the living cell is now a reality. *Nat Chem Biol* 8:879–886.

Osterman N (2010) TweezPal - Optical tweezers analysis and calibration software. *Comput Phys Commun* 181:1911–1916.

Piazzesi G, Reconditi M, Linari M, Lucii L, Sun YB, Narayanan T, Boesecke P, Lombardi V, Irving M (2002) Mechanism of force generation by myosin heads in skeletal muscle. *Nature* 415:659–662.

Pralle A, Prummer M, Florin EL, Stelzer EH, Horber JK (1999) Three-dimensional high-resolution particle tracking for optical tweezers by forward scattered light. *Microsc Res Tech* 44:378–386.

Pyrpassopoulos S, Shuman H, Ostap EM (2010) Single-Molecule adhesion forces and attachment lifetimes of myosin-I phosphoinositide interactions. *Biophys J* 99:3916–3922.

Rai AK, Rai A, Ramaiya AJ, Jha R, Mallik R (2013) Molecular adaptations allow dynein to generate large collective forces inside cells. *Cell* 152:172–182.

Reconditi M, Linari M, Lucii L, Stewart A, Sun YB, Boesecke P, Narayanan T, et al. (2004) The myosin motor in muscle generates a smaller and slower working stroke at higher load. *Nature* 428:578–581.

Ren KF, Grehan G, Gouesbet G (1996) Prediction of reverse radiation pressure by generalized Lorenz-Mie theory. *Appl Opt* 35:2702–2710.

Ren Y, Effler JC, Norstrom M, Luo T, Firtel RA, Iglesias PA, Rock RS, Robinson DN (2009) Mechanosensing through cooperative interactions between myosin II and the actin crosslinker cortexillin I. *Curr Biol* 19:1421–1428.

Rief M, Rock RS, Mehta AD, Mooseker MS, Cheney RE, Spudich JA (2000) Myosin-V stepping kinetics: A molecular model for processivity. *Proc Natl Acad Sci U S A* 97:9482–9486.

Rock RS, Rice SE, Wells AL, Purcell TJ, Spudich JA, Sweeney HL (2001) Myosin VI is a processive motor with a large step size. *Proc Natl Acad Sci U S A* 98:13655–13659.

Rognoni L, Stigler J, Pelz B, Ylanne J, Rief M (2012) Dynamic force sensing of filamin revealed in single-Molecule experiments. *Proc Natl Acad Sci U S A* 109:19679–19684.

Sasaki K, Masanori M, Misawa H, Kitamura N, Masuhara H (1991) Pattern formation and flow control of fine particles by laser-scanning micromanipulation. *Opt Lett* 16:1463–1465.

Skinner GM, Baumann CG, Quinn DM, Molloy JE, Hoggett JG (2004) Promoter binding, initiation, and elongation by bacteriophage T7 RNA polymerase. *A single-molecule view of the transcription cycle. J Biol Chem* 279:3239–3244.

Smith DA (1998) Direct tests of muscle cross-bridge theories: Predictions of a Brownian dumbbell model for position-dependent cross-bridge lifetimes and step sizes with an optically trapped actin filament. *Biophys J* 75:2996–3007.

Smith DA, Steffen W, Simmons RM, Sleep J (2001a) Hidden-Markov methods for the analysis of single-molecule actomyosin displacement data: The variance-Hidden-Markov method. *Biophys J* 81:2795–2816.

Smith DE, Tans SJ, Smith SB, Grimes S, Anderson DL, Bustamante C (2001b) The bacteriophage phi 29 portal motor can package DNA against a large internal force. *Nature* 413:748–752.

Smith SB, Cui Y, Bustamante C (1996) Overstretching B-DNA: The elastic response of individual double-stranded and single-stranded DNA molecules. *Science* 271:795–799.

Smith SP, Bhalotra SR, Brody AL, Brown BL, Boyda EK, Prentiss M (1999) Inexpensive optical tweezers for undergraduate laboratories. *Am J Phys* 67:26–35.

Steffen W, Smith D, Simmons R, Sleep J (2001) Mapping the actin filament with myosin. *Proc Natl Acad Sci U S A* 98:14949–14954.

Stigler J, Ziegler F, Gieseke A, Gebhardt JC, Rief M (2011) The complex folding network of single calmodulin molecules. *Science* 334:512–516.

Svoboda K, Block SM (1994) Force and velocity measured for single kinesin molecules. *Cell* 77:773–784.

Svoboda K, Schmidt CF, Schnapp BJ, Block SM (1993) Direct observation of kinesin stepping by optical trapping interferometry. *Nature* 365:721–727.

Takagi Y, Homsher EE, Goldman YE, Shuman H (2006) Force generation in single conventional actomyosin complexes under high dynamic load. *Biophys J* 90:1295–1307.

Uemura S, Higuchi H, Olivares AO, De La Cruz EM, Ishiwata S (2004) Mechanochemical coupling of two substeps in a single myosin V motor. *Nat Struct Mol Biol* 11:877–883.

Valentine MT, Guydosh NR, Gutierrez-Medina B, Fehr AN, Andreasson JO, Block SM (2008) Precision steering of an optical trap by electro-optic deflection. *Opt Lett* 33:599–601.

van Mameren J, Gross P, Farge G, Hooijman P, Modesti M, Falkenberg M, Wuite GJ, Peterman EJ (2009) Unraveling the structure of DNA during overstretching by using multicolor, single-molecule fluorescence imaging. *Proc Natl Acad Sci U S A* 106:18231–18236.

Veigel C, Coluccio LM, Jontes JD, Sparrow JC, Milligan RA, Molloy JE (1999) The motor protein myosin-I produces its working stroke in two steps. *Nature* 398:530–533.

Veigel C, Molloy JE, Schmitz S, Kendrick-Jones J (2003) Load-dependent kinetics of force production by smooth muscle myosin measured with optical tweezers. *Nat Cell Biol* 5:980–986.

Veigel C, Schmidt CF (2011) Moving into the cell: Single-molecule studies of molecular motors in complex environments. *Nat Rev Mol Cell Biol* 12:163–176.

Veigel C, Schmitz S, Wang F, Sellers JR (2005) Load-dependent kinetics of myosin-V can explain its high processivity. *Nat Cell Biol* 7:861–869.

Veigel C, Wang F, Bartoo ML, Sellers JR, Molloy JE (2002) The gated gait of the processive molecular motor, myosin V. *Nat Cell Biol* 4:59–65.

Visscher K, Gross SP, Block SM (1996) Construction of multiple-beam optical traps with nanometer-resolution position sensing. *IEEE J Sel Top Quant Electron* 2:1066–1076.

Visscher K, Schnitzer MJ, Block SM (1999) Single kinesin molecules studied with a molecular force clamp. *Nature* 400:184–189.

Walter WJ, Brenner B, Steffen W (2010) Cytoplasmic dynein is not a conventional processive motor. *J Struct Biol* 170:266–269.

Walter WJ, Koonce MP, Brenner B, Steffen W (2012) Two independent switches regulate cytoplasmic dynein's processivity and directionality. *Proc Natl Acad Sci U S A* 109:5289–5293.

Wang MD, Schnitzer MJ, Yin H, Landick R, Gelles J, Block SM (1998) Force and velocity measured for single molecules of RNA polymerase. *Science* 282:902–907.

Wang Y, Botvinick EL, Zhao Y, Berns MW, Usami S, Tsien RY, Chien S (2005) Visualizing the mechanical activation of Src. *Nature* 434:1040–1045.

Wen JD, Lancaster L, Hodges C, Zeri AC, Yoshimura SH, Noller HF, Bustamante C, Tinoco I (2008) Following translation by single ribosomes one codon at a time. *Nature* 452:598–603.

Woodside MT, Anthony PC, Behnke-Parks WM, Larizadeh K, Herschlag D, Block SM (2006) Direct measurement of the full, sequence-dependent folding landscape of a nucleic acid. *Science* 314:1001–1004.

Wozniak MA, Chen CS (2009) Mechanotransduction in development: A growing role for contractility. *Nat Rev Mol Cell Biol* 10:34–43.

Wright WH, Sonek GJ, Berns MW (1994) Parametric study of the forces on microspheres held by optical tweezers. *Appl Opt* 33:1735–1748.

Wuite GJ, Smith SB, Young M, Keller D, Bustamante C (2000) Single-molecule studies of the effect of template tension on T7 DNA polymerase activity. *Nature* 404:103–106.

Yin H, Wang MD, Svoboda K, Landick R, Block SM, Gelles J (1995) Transcription against an applied force. *Science* 270:1653–1657.

Zhou R, Kozlov AG, Roy R, Zhang J, Korolev S, Lohman TM, Ha T (2011) SSB functions as a sliding platform that migrates on DNA via reptation. *Cell* 146:222–232.

6 Nanofluidic Transport and Sensing in Biological and Artificial Nanopores

Aleksandr Noy
Lawrence Livermore National Laboratory
Livermore, California
and
University of California, Merced
Merced, California

Meni Wanunu
Northeastern University
Boston, Massachusetts

Contents

6.1 Introduction: Fluid and Ion Flow at the Nanometer Scale in Materials Science and Biology

The second part of the twentieth century witnessed a revolution in materials science that allowed researchers to fabricate and control complex structures with precision that was approaching the scales that were previously encountered only in sophisticated biological systems. This *nanotechnology* revolution pushed scientists to reassess conventional understanding of the nature of mass transport, as well as the physical laws that govern transport behavior and selectivity in these structures (Noy et al. 2007; Schoch et al. 2008; Bocquet and Charlaix 2010). This reassessment, in turn, produced two new and vibrant research fields: nanofluidics and nanopore-based sensing.

Nanopores have traditionally been associated with biological systems where different types of channels that dot the cellular membrane provide some of the most fascinating examples of transport control on a nanometer scale. Ion channels achieve remarkable selectivity, with distinct channels existing for passing sodium, potassium, and chloride ions, the feat unmatched by conventional membranes that cannot discriminate between species with close physical size. On the other side of the biological pore size spectrum, the nucleopore complex achieves an incredible feat of transporting rather large molecules, while still blocking passage of much smaller species. Other types of channels transport water or other uncharged organic molecules without much friction, and with high selectivity. The key to understanding the mechanisms that allow biological pores to achieve this performance with a limited energy budget and under very benign environmental conditions is the realization that they mastered the unique physics of transport at the nanoscale (Box 6.1).

In this chapter, we start by briefly surveying the key physical parameters that are useful for understanding nanofluidic transport, then we describe the main types and structural motives of biological and artificial nanopores, survey some of the resent research on biomimetic nanopores, and, finally, describe the

Box 6.1 What Makes Mass Transport in Channels That Approach Nanometer-Scale Dimensions Special?

Conventional macroscopic-scale descriptions start to fail when we can no longer ignore the contributions of the fluctuations to the overall behavior of the system. For a water droplet to have fluctuations of less than 1%, it needs to contain more than 10,000 particles, which corresponds to a sphere of about 6 nm diameter. This quick estimate tells us that any time water is confined at this scale, we can expect behavior that could be significantly different from the bulk behavior predicted by the classical Navier-Stokes equation.

Another key realization is that when the pore diameter falls below that 6 nm threshold, the channel contains only ca. 20–25 water molecules, and for the pore sizes below 1 nm, the channel starts approaching single-file water arrangements. At these scales, conventional physics parameters, such as solution viscosity, stop being sufficient, and we need to look at the explicit arrangement of hydrogen bonds between water molecules. We also no longer can ignore the details of ion solvation shell arrangement, and cannot completely rely on just the double-layer theory formalism to describe the electrostatic phenomena.

principles and examples of nanopore-based sensing—an approach that promises to revolutionize some areas of medical diagnostics in the coming years.

6.2 Characteristic Physical Dimensions Important for Nanofluidic Systems

Analysis of the key characteristic dimensions of the main interactions that we can encounter in the nanoporous systems also reinforces this notion. Electrostatic interactions between ions and of the ions with the charged walls of surfaces and nanopores define a number of nanofluidic phenomena such as channel selectivity, channel conductance, and electro-osmotic transport, and we can identify several key scales that define the realm of nanofluidics (Gravelle 2016). The first key physical scale that characterizes electrostatic interactions is the Bjerrum length, l_B, that is defined as the length at which the energy of electrostatic interactions between two monovalent ions is equal to the thermal energy k_BT:

$$l_B = \frac{e^2}{4\pi\varepsilon k_B T}$$

where e is the elementary charge, and ε is the dielectric permittivity of the medium. For typical monovalent ions (e.g., K^+ and Cl^-) in water, the l_B value is equal to 0.7 nm, and defines the dimensions at which the hydration of the ions becomes an important factor. Another important scale is defined by the Debye length, λ_D, which defines the characteristic length at which the electrostatic interactions can be considered screened in an electrolyte solution, as well as the characteristic dimensions of a layer that builds up near stationary charges (such as charged surfaces) in solution:

$$\lambda_D = \frac{1}{\sqrt{8\pi l_B C_0}}$$

where C_0 is the ionic species concentration in the solution. In physiologic conditions (at ca. 150 mM salt concentration), the value of λ_D is equal to 0.7 nm (incidentally, very close to the value of the Bjerrum length). Debye length value drops down to 0.3 nm in 1M salt solutions that are commonly used for nanopore electrical conductance measurements. One implication of this dependence is that electrostatic effects play a relatively small role in determining ion transport through moderately sized (diameter above 1 nm) nanopores at high salt (1M and above) conditions, yet start to play a much more important role when the salt concentration drops to physiologic levels or when the channel dimensions drop below 1 nm.

Another relevant yet rarely used parameter for describing electrophoretic transport in nanopores is the Dukhin length, l_{Du}, which defines the channel size at which surface conductance from the ions bound in the Gouy-Chapman layer near a charged nanopore wall is comparable to the conductance through the bulk ions in the middle of the channel. Thus, channels with diameters above l_{Du} behave somewhat like macroscopic channels, where electrophoretic transport is dominated by the bulk transport through channel interior, and in channels with diameters below l_{Du}, electrophoretic transport becomes surface dominated, leading, for example, to different scaling of conductance versus channel size. For a typical charged surface, such as SiO_2, the value of l_{Du} is about 0.5 nm at 1M salt concentration and ca. 3 nm at physiologic level (150 mM), again underscoring the conclusions stated in the preceding paragraph.

The final characteristic length that is relevant to nanofluidics is the slip length, l_{sl}, that defines the interactions of the fluid flowing through the channel with the channel surface. Strictly defined, l_{sl} is the depth inside the solid where the tangential fluid velocity component vanishes, and can be thought conceptually as the measure of the friction the fluid experiences at the wall: the more slippery the surface is, the larger is the value of the slip length, and it reaches infinity for a perfectly slippery wall that does not interact with the fluid molecules. Since the typical value of the slip lengths for simple liquids at surfaces are on the order of 20–80 nm (Bocquet and Barrat 2007), it is clear that conventional hydrodynamic flow descriptions need to undergo serious modification to be able to describe fluid flow in nanofluidic channels.

6.3 Biological and Biomimetic Nanopores

6.3.1 Protein Nanopores

Biological proteins provided some of the first and still the most astonishingly elaborate examples of nanopore design (Aidley and Stanfield 1996). Cells take advantage of compartmentalization as a way to separate important metabolic processes and insulate their interior from the environment. Consequently, they have evolved an arsenal of proteins that mediate transport across the membranes that separate the compartments (Alberts et al. 2007). Typical sizes of these transmembrane pores range from 0.3 nm aquaporin pores, which allow only water molecules to squeeze through (Agre et al. 1998; Murata et al. 2000), to the giant pores of the nuclear pore complex, which can transport particles as large as 25 nm (Adam 2001). A prototypical membrane protein pore, such as α-hemolysin from *Staphylococcus aureus* (Figure 6.1a), which is commonly used in nanopore sensing applications, arranges several β-sheets into a highly symmetrical β-barrel domain that anchors in the lipid membrane and forms the walls of the pore (the *barrel-stave* motif) (Song et al. 1996). This motif is ubiquitous in the membrane pore structure: another key protein nanopore used in pore-sensing applications, MspA (Figure 6.1b), also contains a β-barrel (Faller et al. 2004). Smaller pores, such as aquaporins (Figure 6.1c) often use a different arrangement based on α-helical architectures to form the membrane-spanning pore domains (the *coiled-coil* motif). Fully synthetic nanopores, such as carbon nanotubes (Figure 6.1d), are built on a very different structural paradigm and be discussed later in the chapter.

These two motifs, coupled with the astonishingly rich structural diversity of the peptide architecture, allowed nature to precision-engineer transmembrane nanopores with the side-chains of the main structural elements contributing charged groups, forming in-channel constrictions, and creating precision selectivity filters that give the channels unparalleled specificity. Notably, protein pores exhibit a significant diversity of their selectivity mechanisms, which include

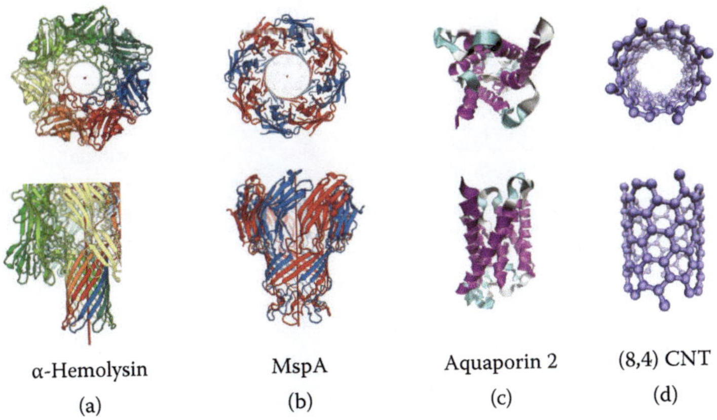

Figure 6.1 Biological and biomimetic nanopores. (a) α-hemolysin protein. (b) MspA protein. (c) Human aquaporin (AQP) protein. (d) 0.8 nm diameter (8,4) carbon nanotube. The top row shows the top view of the nanopore structure; the bottom row shows the side view.

charge-based and size-based rejection, as well as specific mechanisms based on recognition binding and other triggered structural transitions that block the channels and open and close protein lids.

An ion or a macromolecule moving through a biological nanopore encounters several distinct regions while transiting the pore. For example, a macromolecule translocating through an α-hemolysin pore (Figure 6.2) from the cis-side encounters in sequence a 2.6 nm cis-opening, an inner "vestibule" cavity, and a 1.3 nm constriction (Song et al. 1996). Upon passing through that narrow part of the channel, the molecule slips through the 2.2 nm β-barrel region that finally lets the molecule out of the pore through the trans-side opening. Significantly, it is usually the narrow constricted section of this path that determines the dynamics of the macromolecule translocation and constitutes the sensing region of the pore, which is responsible for the majority readout interactions. The MspA pore (Figure 6.1b), which is also used for nanopore sensing, has a similar structure with a funnel-like channel ending in a constriction.

A water transporting protein, aquaporin (AQP) has a somewhat different structure in which the protein helices form a very tight 0.3 nm diameter channel that squeezes water molecules passing through that channel down to a single-file chain configuration (Murata et al. 2000). The very small diameter of the AQP pore gives it 100% ion rejection because the energetic cost of compete desolvation that is necessary for ions to pass through that channel is too high. AQPs also display an unusual and rather clever mechanism of blocking proton transport, which is necessary for the cells to avoid excessive proton leakage through the membrane and the resulting depolarization (Ilan et al. 2004). Since protons can translocate

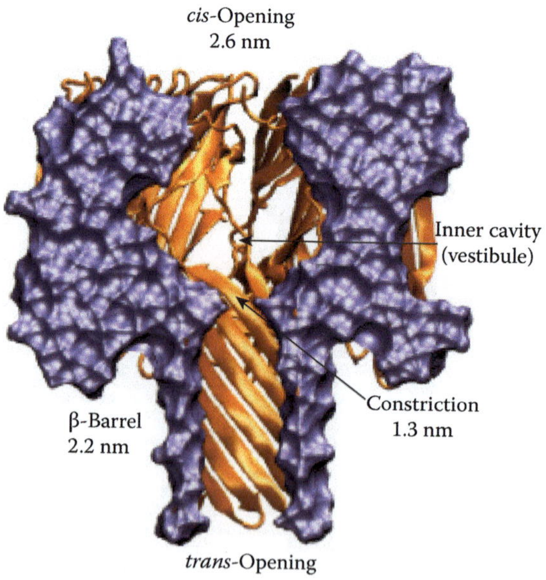

Figure 6.2 Structure of α-hemolysin pore. A cutaway showing the major elements and corresponding sizes of the α-hemolysin protein pore.

effectively along the hydrogen-bonded chain of water molecules via the Grotthuss mechanism, which involves charge hopping along the chain of hydrogen bonds, the AQP channel features partial charge arrangement in the middle of the channel at the point close to the tightest constriction, which makes a water molecule flip as it passes by these charged groups. This flip motion disrupts the continuous hydrogen-bonded chain of water molecules and blocks the Grotthuss proton transport. Since the only other available mechanism of proton transport, standard (Fickian) diffusion of hydronium ion (H_3O^+) is much slower than Grotthuss transport and involves squeezing a bigger species through the constriction region, it makes the proton rejection effectively complete. These properties make AQPs some of the most dramatic examples of separation selectivity that can be achieved in biological nanopores and motivate the ongoing efforts to develop biomimetic desalination membranes based on AQP protein arrays.

6.3.2 Synthetic Nanopore Channels

The beauty and efficiency of natural membrane nanopore structures have been inspiring chemists and biochemists to create synthetic mimics for a long time (Fyles 2007). Broadly, the examples of synthetic channels reported in literature can be divided into two classes. The first class, tubular channels, has structures that are akin to the gramicidin channel, in which the transmembrane passage is formed by stacking of well-defined tubular subunits (Figure 6.3). In the second, aggregate channels are formed by the coming together of several membrane-spanning subunits; this mechanism is similar to the formation of the amphotericin channel (Box 6.2, Figure 6.3).

Box 6.2 Synthetic Membrane Pores

Some examples of synthetic membrane nanopores include tris-macrocycle hydrophiles (Murray et al. 2000) that use three large crown ether subunits to form a transmembrane structure. Another example of a gramicidin-like structure is a ca. 2.5 nm diameter rigid-rod β-barrel formed by octiphenyl derivatives (Matile et al. 2004). The large pore size of these channels makes them well-suited for blocking by large biomolecules, and experiments showed that they could be obstructed by DNA oligomers. Another popular approach for creating synthetic ion channels focuses on cyclic peptides that form a stacked barrel structure (a *peptide nanotube*) in the membrane (Fyles 2007). Aggregate channels typically exhibit less structural precision than the stacked structures, but since they use building blocks that span the membrane (typically two-headed amphiphiles), they can, in principle, give rise to asymmetric structures that could exhibit voltage-gated functionality. Examples of such channels are based on polymacrocycles or acyclic oligoesters, and others draw upon the same octiphenyl scaffold used in gramicidin-like structures. Designs based on the antibiotic squalamine that feature sulfate groups showed good antiviral activity against HIV in addition to the membrane activity (Jing et al. 2004).

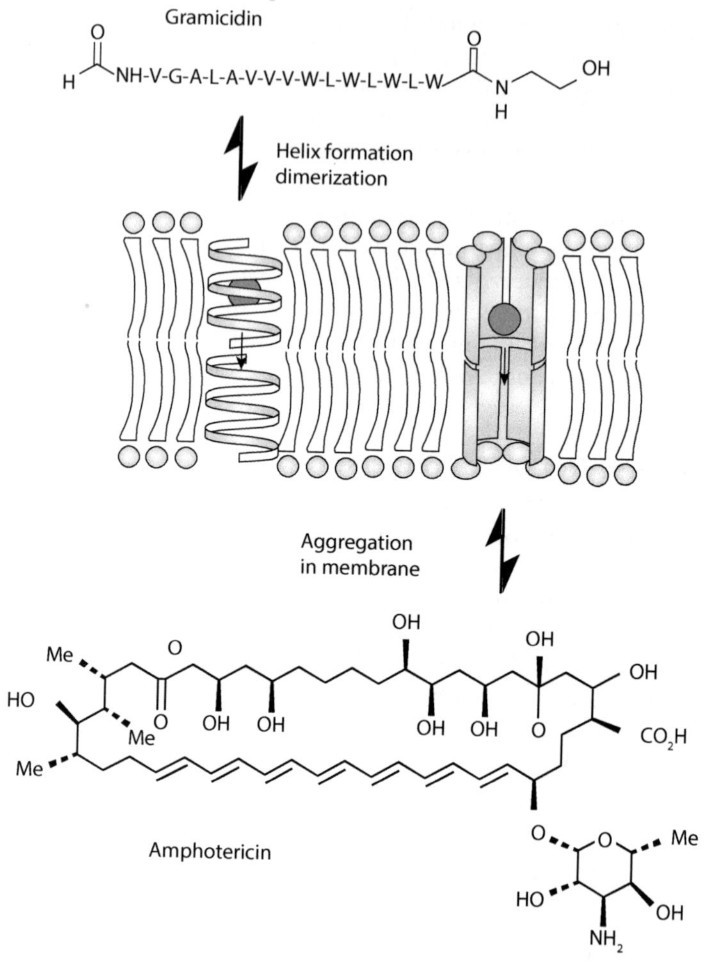

Figure 6.3 Examples of ion channels formed by naturally occurring compounds. The peptide gramicidin forms a head-to-head helical dimer that opens a tubular channel within a bilayer membrane. The antibiotic amphotericin forms an aggregate channel surrounding a membrane pore. (From Fyles, T., *Chem. Soc. Rev.*, 36, 335–347, 2007. With permission.)

Despite the unquestionable successes of the synthetic ion channel design efforts, these molecules remain staggeringly complex, and, more often than not, the design is driven by constraints such as relative synthetic difficulty, which reduces the researcher's ability to tailor these structures to a particular task.

6.3.3 Carbon Nanotube Nanopores

Problems associated with creating highly efficient synthetic membrane nanopores—synthetic difficulties as well as poor transport efficiency of these channels—motivated researchers to try other design paradigms based on channel scaffolds that are rooted in the nanomaterials field. Carbon nanotube porins (CNTPs) (Figure 6.1d)—short segments of carbon nanotubes (Geng et al. 2014)—represent perhaps the simplest pore-forming structure and open up a number of opportunities for studying fundamentals of the nanofluidic transport. We discuss the structure of the CNT porins and some of their transport properties in Section 6.5.2.

6.3.4 Solid-State Nanopores

Solid-state nanopores are nanopores made by carving a hole in a solid-state membrane, typically less than 100 nm thick. The most common nanopore membrane material is a low-stress silicon nitride membrane supported by a silicon chip frame. Methods for fabrication of these nanopores range from wet or dry chemical methods, electron-beam irradiation, ion-beam irradiation, and most recently, active feedback high-voltage dielectric breakdown of the membrane material (Kwok et al. 2014). Since nanopore diameters and thicknesses can be controlled by manipulating the fabrication conditions, various species can be detected, from short DNA molecules to protein aggregates. However, two main challenges must be mentioned, and these are distinguishing features of solid-state nanopores from the other pores discussed in the previous sections. The first is that most commonly, the nanopores are fabricated under vacuum, and then they must be hydrated prior to transport measurements. The second is that no two nanopores fabricated are structurally and chemically identical, which often produces significant heterogeneity from experiment to experiment. Although modern experimental approaches have partially addressed these challenges, these issues are beyond the scope of this chapter, and we refer the reader to the literature (Wanunu 2012).

6.4 Methods to Probe Transport through Nanopores

The very small nanopore size, which is what makes the transport through them so interesting, also creates significant difficulties in designing experimental techniques to detect transport flux. For example, a single AQP pore, which has a significant water permeability of 10^{-13} cm^3/s (de Groot and Grubmüller 2005), would take more than 300 years to transport 1 mg (1 μL) of water, and that estimate assumes ideal conditions that an experiment rarely (or rather never!) achieves. Thus, transport measurements have to rely either on assembling nanopores into arrays (such as in carbon nanotube membranes [Holt et al. 2006]) or packing them into the walls of lipid vesicles and measuring aggregate flux. The last type of measurement (see Figure 6.4 for a typical example) is usually performed using a stop-flow apparatus that can detect fast kinetics typically associated with nanopore transport (Kumar et al. 2007; Barboiu 2012).

The advent of sensitive low noise current amplifiers that can pick up and record currents in the 10^{-12}–10^{-9} A range associated with ion transport through typical nanopores allowed researchers to introduce simple and sensitive techniques for measuring electrophoretic currents through nanopores. A particularly widely used technique—planar lipid bilayer or black lipid membrane (BLM) measurement—uses a two-compartment fluid cell that is separated by

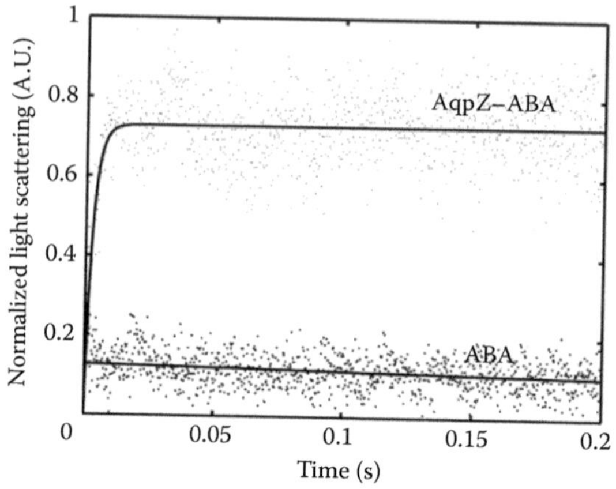

Figure 6.4 Stop-flow measurement of water permeability of water transporter Aquaporin-Z (AqpZ) in polymer vesicles. The data track the light-scattering signal associated with the rapid shrinking of the vesicles that contain AqpZ under osmotic pressure gradient. (From Kumar, M., et al., *Proc. Natl. Acad. Sci.*, 104, 20719–20724, 2007. With permission.)

a thin Teflon partition with a small (ca. 50 μm) orifice. The orifice is sealed by a thin layer of lipid membrane, and a pair of electrochemical electrodes situated in each of the two compartments measures the current between the two compartments. The lipid membrane typically forms a very tight electrical seal (ca. >5 GΩ); thus, incorporation of even a single nanopore channel into the lipid bilayer drops the membrane resistance significantly and produces measurable current response (Figure 6.5) (Wanunu 2012). This setup allows two general modes of measurements (Wanunu 2012): open pore measurement in which the current steps associated with incorporations of individual nanopores into the bilayer characterize individual pore conductance, and the blockade detection mode in which the polymers, such as DNA or PEG, are dragged into the pore using either electrophoretic or electroosmotic flow. Passage of the polymer through the pore orifice produces "resistive pulse" that is detected by the amplifier and can be used to determine the transport characteristics of the polymer.

6.5 Nanofluidics in Carbon Nanotubes and Carbon Nanotube Porins

6.5.1 Carbon Nanotube Membranes

A carbon nanotube forms when a graphene sheet is rolled up into a perfectly seamless cylinder. It is common to characterize the structure of the nanotube by its rolled-up vector (n,m) that defines the position of the matched carbon rings in the rolled-up structure and determines the nanotube's chirality (or helicity) (Dresselhaus and Saito 1998). The roll-up vector fully defines the nanotube morphology, its electronic properties, as well as its diameter.

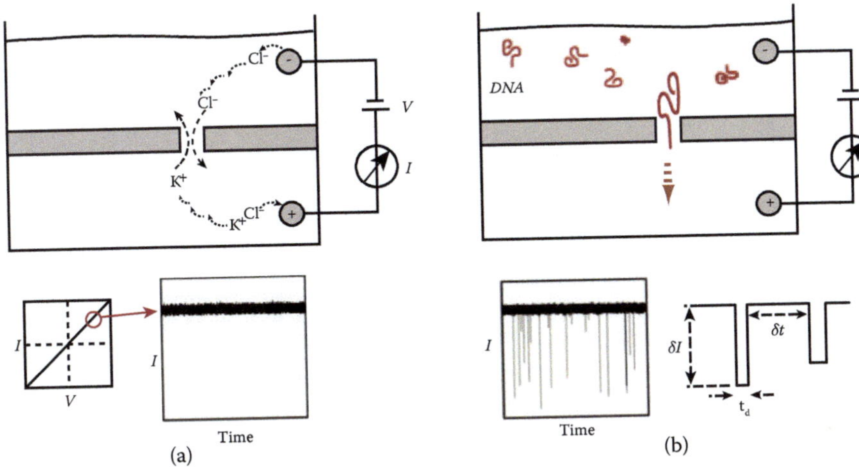

Figure 6.5 Nanopore sensing basics. (a) Application of voltage across a pore triggers electrochemical half-reactions leading to ion migration toward the membrane. Transport of the ions across the pore leads to electric current that is measured using an electrometer. Holding the potential constant results in a steady-state DC current signal, which can be measured at a high bandwidth. (b) When charged biopolymers are added to the chamber, such as DNA molecules, biopolymer molecules diffuse toward the pore and stochastically enter it, producing a measurable "resistive pulse." First-order parameters that can help in characterizing a sample are the dwell time (t_d), the average event amplitude (δI), and the time between successive events (δt). (Adapted from Wanunu, M., *Phys. Life Rev.*, 9, 125–158, 2012.)

A carbon nanotube can have one (as in a single-walled carbon nanotube, SWCNT), or several (as in a multiwall carbon nanotube, MWCNT) concentric graphitic shells (Dai 2002). Both types of nanotubes retain the remarkable structural properties that make the nanotube so attractive for nanofluidic applications: extremely high aspect ratio and small lateral dimensions (a nanotube can easily reach up to several millimeters in length, yet retain a diameter of only a few nanometers), as well as the remarkable atomic scale smoothness and chemical inertness of the nanotube's graphitic walls.

MD simulations first provided a clear indication small-diameter CNTs could be an interesting nanofluidic substrate due to their tendency to form hydrogen-bonded "water wire" structures in which the water molecules form a "train" that runs down the nanopore axis (Hummer et al. 2001; Kalra et al. 2003; Hummer 2007). Interestingly, this structure is highly reminiscent of the wire-like structures observed in the biological channels, such as aquaporins (Sui et al. 2001). The similarity between aquaporin channels and carbon nanotube channels extends further to the inner pore functionality. Similar to the hydrophobic interior of carbon nanotubes, the inner cavity of the aquaporins is lined with hydrophobic residues that facilitate transport through the pore (Murata et al. 2000; Sui et al. 2001). These features are key to the phenomena of nearly frictionless transport of water in nanotubes channels (Kotsalis et al. 2004; Hanasaki and Nakatani 2006; Striolo 2006). Early on, Kalra and Hummer et al. showed that water moves very rapidly through a nanotube channel under osmotic pressure (Kalra et al. 2003), and showed that friction at the channel walls in that system was so low that water transport was no longer governed by the Hagen-Poisselle flow framework, but instead depended mainly on the events at the nanotube entrance and exit (Kalra et al. 2003), with calculated rates of water transport (6 H_2O molecules/ns) comparable or exceeding the transport rates observed in the best biological water transporters (Agre et al. 2001).

These predictions were tested with 1.6 nm inner-diameter double-wall carbon nanotube membranes, made by filling vertical free-standing arrays of CNTs with an inert silicon nitride matrix and then removing top and bottom layers to expose and open the nanotube ends (for details of the structure see the membrane images in Figure 6.6a through e, and the transmission electron microscopy (TEM) characterization data in Figure 6.6f and g). These membranes showed high rates of water transport through the double-wall sub-2-nm CNT membranes using pressure-driven flow (Holt et al. 2006). Similarly, high rates were also observed by Majumder et al. using multiwalled nanotube membranes with larger pore diameters (Majumder et al. 2005). Curiously, the reported enhancements over Hagen-Poiselle flow rates range from two to five orders of magnitude. This astonishingly wide range of enhancements reflects a number of problems that complicate the experiments, such as virtual inability to determine accurate pore density of the nanotube membranes, as well as the generally flawed strategy of comparing the low rates with Hagen-Poisselle flow (Craig et al. 2001; Cottin-Bizonne et al. 2002; Majumder et al. 2005), and underscore the difficulties associated with interpreting the data from nanofluidic experiments.

6.5.2 Carbon Nanotube Porins

The similarities between the transport through carbon nanotubes and biological nanopores raise an intriguing possibility of using the carbon nanotube to create biomimetic ion channels. Noy and coworkers have recently reported such nanostructures—carbon nanotube porins (CNTPs) (Geng et al. 2014)—which consist of short (ca. 10 nm length) fragments of carbon nanotubes wrapped with lipid molecules and inserted into a lipid bilayer. CNTPs are a pedagogically interesting example of a biomimetic nanopore because they illustrate how we can capture many of the same fundamental physical principles that enable biological transport—confinement, hydrophobic channel walls, and electrostatic exclusion—in a controlled synthetic pore scaffold.

The lipid coating is responsible for two important traits of the CNTPs. First, they are soluble, forming a long-term stable dispersion in water; second, this lipid coating gives them the ability to self-insert into the lipid membranes and form well-defined nanopores (Figure 6.7a). Cryogenic TEM characterization (Figure 6.7b) confirms that CNTPs form pores that run nearly perpendicular to the lipid bilayer plane.

In the BLM experiments, the addition of the CNTPs to one of the chambers of the measurement setup (Figure 6.8a) produces unmistakable current steps that indicate rapid CNTP incorporation into the lipid membrane. The average

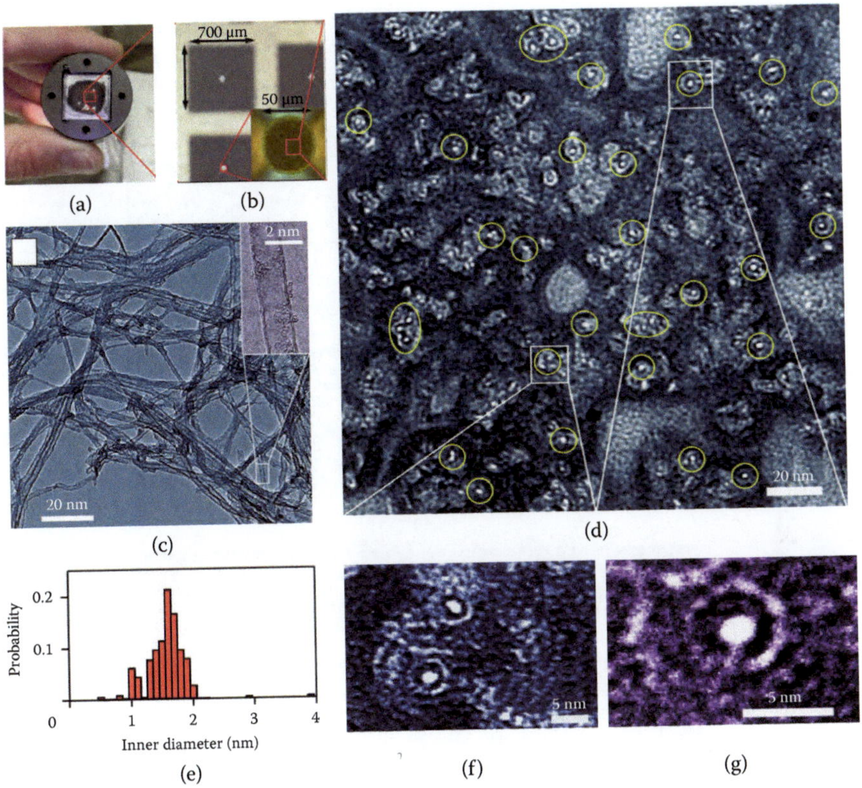

Figure 6.6 Carbon nanotube membranes. (a) A CNT membrane chip in the sample holder. (b and d) Optical micrographs of the regions of the chip that contain CNT membrane windows. (d–g) High-resolution TEM images of the thinned cross sections of the membrane showing sub 2 nm pores. (c and e) HR-TEM characterization of the CNT size: (c) TEM image of the dispersed carbon nanotubes from the vertically aligned array used for membrane fabrication; (b) histogram of inner diameters of carbon nanotubes. (From Holt, J. K., et al., *Science*, 312, 1034–1037, 2006. With permission.)

conductance of the 1.5 nm CNTP of ca. 630 pS is consistent with the typical conductance value range of biological nanopores of comparable size. Some fraction of the 1.5 nm CNTPs also exhibits a fascinating tendency for stochastic gating, which originates in the rare ionic "liquid-to-vapor" transition that exists in a narrow range of nanopores at around 2 nm diameter (Buyukdagli et al. 2010; Buyukdagli et al. 2011). The detailed description of these events is beyond the scope of this chapter.

A significant advantage of the CNTP as a platform for nanofluidic studies is that they not only can be used in single nanopore measurements, but they also permit bulk vesicle-based assays, analogous to the assays that have been used for decades to study transport in membrane proteins. One such assay is the osmotic permeability assay (Figure 6.9) that can probe selectivity of the CNTPs

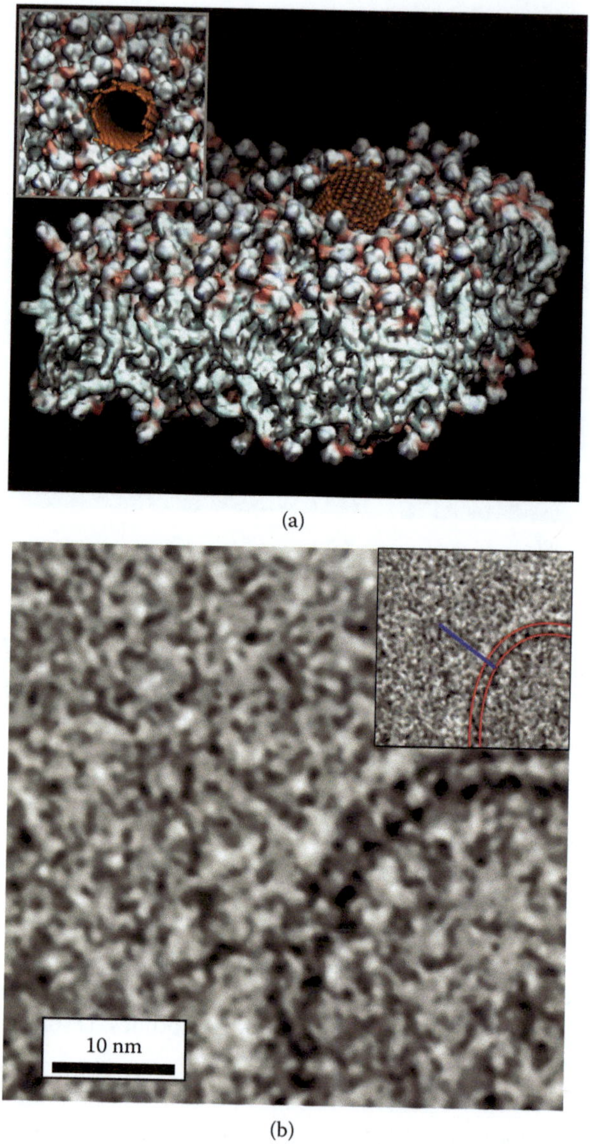

Figure 6.7 Carbon nanotube porins. (a) Snapshot from MD simulation of a CNTP in a lipid bilayer. (b) TEM image of a CNTP inserted into the lipid membrane. Ca. 15° angle CNTP tilt evident in the TEM images most likely represents a compromise between the nanotube's propensity to bury its hydrophobic surface in the lipid bilayer interior and the energetic penalty of disrupting the lipid packing inevitable for larger tilt angles.

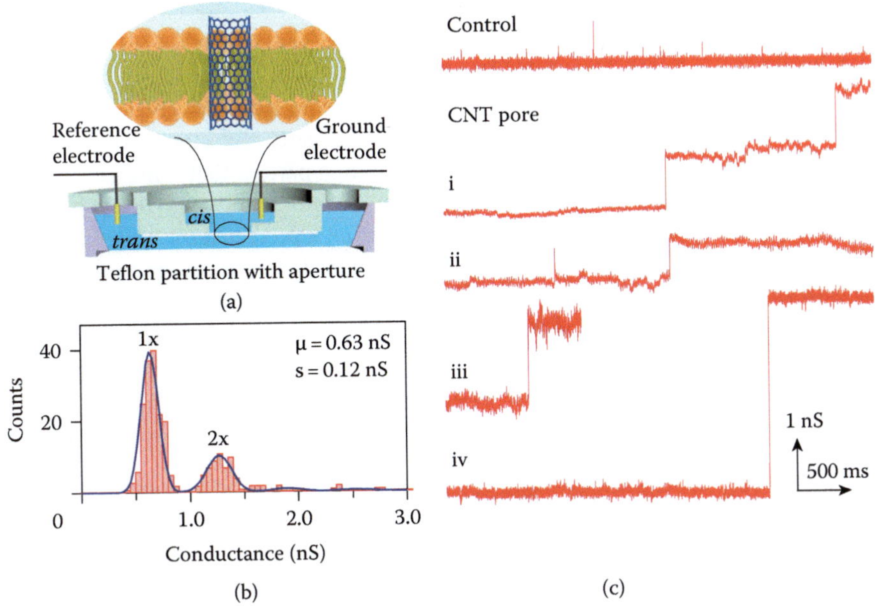

Figure 6.8 Conductance properties of individual CNTPs. (a) The BLM measurement setup. (c) Current traces showing distinctive current steps resulting from CNTP incorporations. (b) Histogram of the magnitude of conductance steps measured for incorporation of CNTPs into lipid bilayers. Blue line represents the fit to two Gaussian peaks representing conductance of one and two CNTPs. (From Geng, J., et al., *Nature*, 514, 612–615, 2014. With permission.)

(Kim et al. 2014); see Box 6.3 for examples of using this assay to elucidate the charge and size selectivity of the CNTPs.

One-dimensional "water wire" structures (which were briefly discussed in Section 6.3.1) also play an important role in the proton transport via Grotthuss charge hopping mechanism in biological structures such as bacteriorhodopsins and cytochromes (Pomes and Roux 1996; Schlichting et al. 2000; Brewer et al. 2001). Not surprisingly, molecular dynamics simulations indicated that carbon nanotube channels should also facilitate proton transport in the direction along the nanotube axis (Dellago et al. 2003). The striking prediction was that the proton mobility in the nanotube should exceed the value in bulk water by an order of magnitude, thus achieving perhaps the ultimate efficiency of the Grotthuss proton transport mechanism.

Recent studies of proton transport through the CNTP interior pores (Figure 6.10) confirmed these predictions (Tunuguntla et al. 2016). Proton transport in 0.8 nm diameter CNTPs was extremely fast, reaching the predicted 10× enhancement over the bulk value, and confirming the critical role water ordering plays for proton transport. These data underscore the reasons why biological proton channels evolved to create high water confinement, and also provide perhaps the clearest direct confirmation yet for the Grotthuss mechanism of proton transport through water.

Another confirmation of the importance of water confinement in transport through nanopores comes from the study of the proton transport efficiency in wider 1.5 nm diameter CNTPs, which fit more than one water molecule across the channel. These wider channels show a hydrogen bonding pattern that is not very distinct from the bulk water hydrogen bonding pattern, and not surprisingly, they show proton diffusion efficiency that is very close to the value predicted for bulk water. This comparison indicates that narrow nanopores can facilitate efficient proton transport by directing the proton hopping along the pore direction, as opposed to wider channels where the protons can spend a long time hopping in the "sideways" directions. Recently Noy and Wanunu reported very high water fluxes and high ion selectivity in narrow diameter CNTPs, which indicates that the separation performance of these pores is suitable for desalination applications (Tunuguntla 2017).

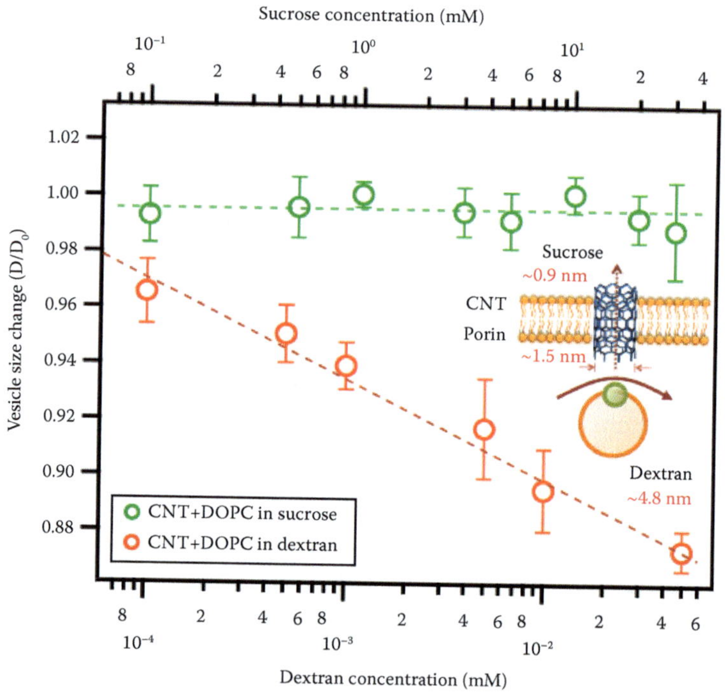

(a)

Figure 6.9 Transport selectivity in carbon nanotube porins. (a) Plot of the size change of vesicles containing CNT porins after exposure to solution containing different concentrations of sucrose and dextran. Inset shows a schematic of CNT porin in lipid bilayer and hydrodynamic radii of dextran and sucrose. Dashed lines are guides to the eye. Only dextran solutions induce measurable changes in the vesicle size, indicating that CNTPs are impermeable to large molecules.

(Continued)

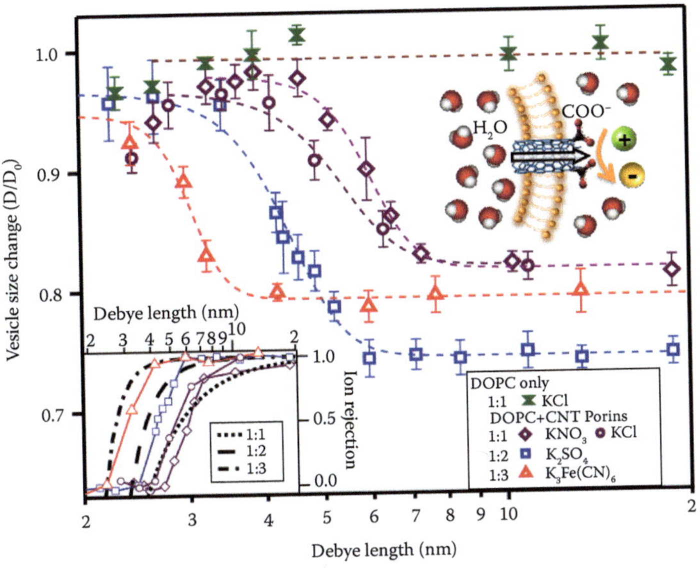

(b)

Figure 6.9 (Continued) Transport selectivity in carbon nanotube porins. (b) Plot of the vesicle size change as a function of the solution Debye length after exposure to the different 1:1, 1:2, and 1:3 electrolyte solutions of different concentrations. Electrolytes used for each experiment are indicated on the legend. Dotted lines indicate a fit to a sigmoidal function. Schematic on the graph illustrates how charged tips of the CNT porins reject ion passage through the nanotube pores. At low ion concentrations (high Debye length values), the charges on the nanotube reject ion passage through the pores, and thus allow the osmotic pressure to squeeze water out of the vesicles. Enhanced screening at higher salt concentration drives down the ion rejection, and thus diminishes the efficiency of the osmotic "squeeze" process. (Inset) Comparison of the CNTP ion rejection data (markers and solid lines, same as indicated on the main figure legend) with the predictions of the Donnan model (dashed and dotted lines). Good agreement between the experimental data and theory prediction illustrates the key role that charge repulsion plays in the mechanism of ion rejection. (From Kim, K., et al., *Nano Lett.*, 14, 7051–7056, 2014. With permission.)

CNTPs represent a versatile lipid membrane nanopore system that reproduces a surprisingly complete range of the simplest transport and selectivity mechanisms used by the biological membrane channels. Their robustness and ease of manufacture are vastly superior to their biological counterparts; however, the selectivity of the nanopores is still very far from the standards displayed by their biological analogs. One of the hopes for the future of the CNTP system lies in the further development of this promising platform, which would allow researchers to incorporate much more sophisticated selectivity mechanisms into the CNTPs and allow them to play a more important role in future biomimetic structures and sensing systems.

Box 6.3 Size and Charge Selectivity of CNTPs Probed with the Osmotic Permeability Assays

Exposure of the vesicles with CNTP nanopores incorporated into the vesicle walls to an osmotic gradient pushes water out of the vesicles and leads to their shrinkage, which produces detectable changes in the light-scattering intensity. Notably, to create a durable osmotic pressure gradient, the osmolyte cannot travel through the CNTP pore, otherwise diffusion into the vesicle interior would quickly dissipate the osmotic gradient. The comparison between the CNTP vesicle response to the osmotic challenge by two different molecular weight sugars, sucrose and dextran (Figure 6.9a), illustrates this concept and shows that CNTP are able to discriminate between the uncharged molecules based on the molecule size.

The technique can also be adapted to probe the mechanism of ionic selectivity of CNTPs (Figure 6.9). In this case, the vesicle response to the osmotic pressure challenge is determined by the ion's interactions with the charged groups at the nanotube porin entrance, which can be tuned by changing the overall ionic strength of the solution, essentially manipulating the Debye length. In this case, the electrostatic potential of charged groups immobilized on the mouth of the CNTP nanopore act as a gate that prevents the ions from entering the nanotube. Changing the electrolyte Debye length modulates the overlap of the electrostatic field from these groups, effectively opening and closing the pore. Indeed, lipid vesicles with CNTPs embedded in their walls respond very differently to osmotic challenges by electrolyte solutions of different concentrations (Figure 6.9b). At high salt concentrations (e.g., smaller values of Debye length), the electrostatic gate remained open and the osmotic gradient got equalized rather quickly. In contrast, when the similar vesicles were challenged by low ionic strength salt solution, they showed much stronger size change. Interestingly, the smooth transition from this "high-rejection" behavior at low ionic strengths to a "low-rejection" response at higher ionic strength maps very well on the pore selectivity curve predicted by the Donnan model of ion selectivity in a charged membrane nanopore (Figure 6.9b, inset) (Kim et al. 2014).

6.6 Nanopore-Based Sensing: Nucleic Acid Transport through Nanopores

6.6.1 Origins of Nanopore-Based DNA Sequencing

In their seminal 1996 work (Kasianowicz et al. 1996), Kasianowicz, Brandin, Branton, and Deamer demonstrated that single-stranded nucleic acids (in the particular case, a homopolymer of RNA, polyU) can be electrophoretically driven through a α-hemolysin channel. Three key observations were made: First, the rate of spikes depended on the bulk concentration of the RNA sample. Second, three

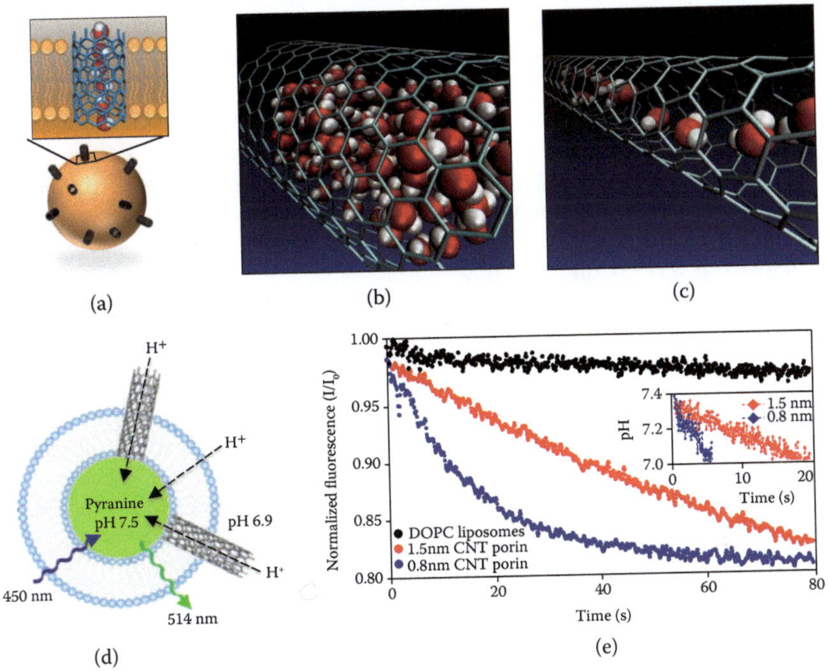

Porin	Pore inner diameter (nm)	Pore length (nm)	H^+ conductance (nS/pore)	H^+ conductivity (S/cm)	Diffusion constant (A^2/ps)	
			Measured	Measured	Measured	Calculated
1.5 nm CNT	1.4 ± 0.2	13.4 ± 5.4	$1.80 \times 10^{-7} \pm 6.9 \times 10^{-8}$	$1.70 \times 10^{-8} \pm 6.5 \times 10^{-09}$	0.76 ± 0.18	–
0.8 nm CNT	0.7 ± 0.1	10.6 ± 0.9	$3.30 \times 10^{-7} \pm 7.2 \times 10^{-8}$	$9.20 \times 10^{-8} \pm 1.9 \times 10^{-9}$	4.20 ± 0.91	3.86*
Gramicidin A	0.4	5	1.20×10^{-7}	4.76×10^{-8}	0.35	–
Nafion™	–	–	2.1×10^{-8}**	1.8×10^{-9}**	–	–
Water (bulk)	–	–	–	–	–	0.40*

* Values reported by ref. (17); ** Assumes a 10 nm pore with a 0.7 nm diameter filled with Nafion and scaled to pH=7 conditions (12).

(f)

Figure 6.10 Proton transport in CNTPs. (a) CNTPs inserted into a liposome. (b and c) Molecular models showing water molecules in the inner pores of two CNTPs of different diameters: 1.5 nm (b), and 0.8 nm (c). (d) The proton conductance measurement. (e) Normalized pyranine dye fluorescence intensity changes (I/I0) as a function of time. Inset: Magnified initial region of pH gradient dissipation kinetics for 0.8 and 1.5 nm diameter CNTPs in DOPC liposomes. Solid lines are a linear fit to the data in pH range from 7.5 to 6.9. (From Tunuguntla, R. H., et al., *Nat. Nanotechnol.*, 11, 639–644, 2016. With permission.)

types of spikes were present, segregated into three populations of dwell times (short, intermediate, and long). Third, the dwell times of the intermediate and long spikes were inversely proportional to the voltage applied, while the short dwell time population was insensitive to applied voltage. These observations suggested that the RNA homopolymers threaded through the nanopores, with the intermediate and long populations representing transport of the molecule in different orientations, as later showed by clever experiments from Mathé and coworkers (Mathé et al. 2005). The manuscript concluded with an extraordinary statement (p. 13773), quoted below:

> The realization that single channel measurements can be used to determine the lengths of individual RNA or DNA chains suggests that other characteristics of a polynucleotide could also be measured as the molecule traverses the channel. In principle, a single purine or pyrimidine nucleotide passing through the limiting aperture of a channel should block the ionic current in a manner that reflects the molecular size and chemical properties of each nucleotide in a polynucleotide. If so, it may be possible to use single channel recordings to directly determine the sequence of a polynucleotide.

This statement captured the essence of nanopore research over the next two decades, which aimed to sequence DNA using nanopores. In 1996, capillary-based Sanger DNA sequencing was the prominent method of choice, and "worker bee" researchers from around the world were in full swing to achieve a grand feat: sequencing of the entire human genome, a project led by the National Human Genome Research Institute (NHGRI) of the National Institutes of Health (NIH). Sequencing at the time, and in the years to follow, had been performed by reading a few hundred bases of fragmented DNA at a time, and then stitching together these sequence reads to longer assemblies, a method that indeed has proven itself, despite its limitations and errors, to provide the first glimpses into the human genome scaffold in 2001 (Lander 2001). The idea that instead of making many copies of short DNA fragments, a nanopore could read a single long molecule that is tens of thousands of bases long or more, was revolutionary. However, as data presented in Figure 6.11 may already foreshadow to the perspicacious reader, the presence of a complex distribution of dwell times, three populations, broad in dwell times, highlights some of the challenges that were to be overcome in order to achieve DNA motion control for sequence readout.

6.6.2 Voltage-Driven DNA Transport

Following the work of Kasianowicz and coworkers, Meller and Branton studied in detail the transport kinetics of different DNA homopolymers through α-hemolysin. Of particular note was the shape of the distributions, which was not perfectly Gaussian in nature, but rather exhibited a fast rise at short times and a tail of varying decay lengths that hinted at interactions of the molecules within the pore (see Figure 6.12a) (Meller et al. 2000). For a distribution of this type, the mean time is longer than the most probable time. Thus, the homopurine poly(dA)

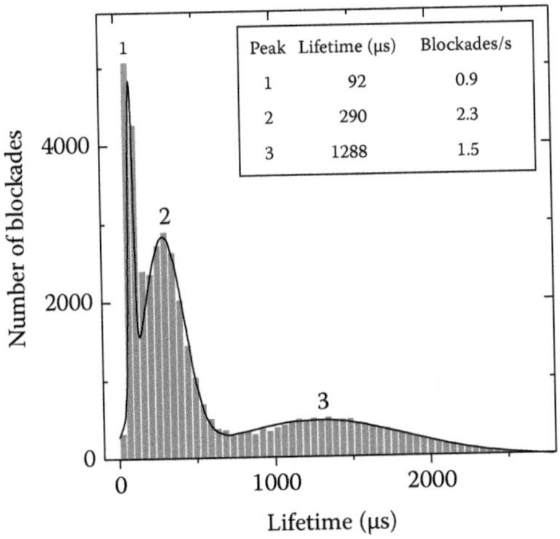

Figure 6.11 Distribution of electrophoretic current spike durations recorded during RNA polyU_{210} translocations through α-hemolysin, showing a polydisperse, multipopulation distribution. (From Kasianowicz, J. J., et al., *Proc. Natl. Acad. Sci.*, 93, 13770–13773, 1996. With permission.)

interacts more with the pore than the polypyrimidine poly(dC), and these interaction differences result in different peak pulse durations (t_P) and tail shapes. The behavior is somewhat similar for the transport of long double-stranded DNA (dsDNA) through 10 nm diameter SiO_2 pores, as revealed by Storm and coworkers (see Figure 6.12b) (Storm et al. 2005).

More complex pulse duration distributions were obtained for dsDNA interactions with 4 nm diameter silicon nitride pores (twice the cross-sectional diameter of dsDNA). In the limit of small pores, a broad-tailed distribution is observed that best fits a two-exponential decay. Moreover, the relative fraction of events in each population is length dependent, and the longer the DNA molecule, the more events in the long timescale are found.

Recently, experiments that probed the velocity of double-stranded DNA transport were carried out (Singer et al. 2012; Plesa et al. 2015; Bell and Keyser 2016). In these experiments, DNA was labeled by a bulky group at a sequence-specific site (either a peptide-nucleic acid or an additional DNA strand), and ion current fluctuations were measured to discriminate DNA position along the translocation coordinate. For example, in experiments by Bell and Keyser, a DNA construct made up of sausage-like links of DNA origami was constructed (see Figure 6.13a), wherein at every few kilobase intervals, a fatter construct of DNA was assembled by designing an origami-like loop that would provide an ion-current contrast when entering the pore. The series of spikes within the DNA translocation event reports on the translocation coordinate of the DNA molecule in the pore. By quantifying the interspike durations in the nanopore transport experiment (Figure 6.13b and c), the researchers were able to extract velocity

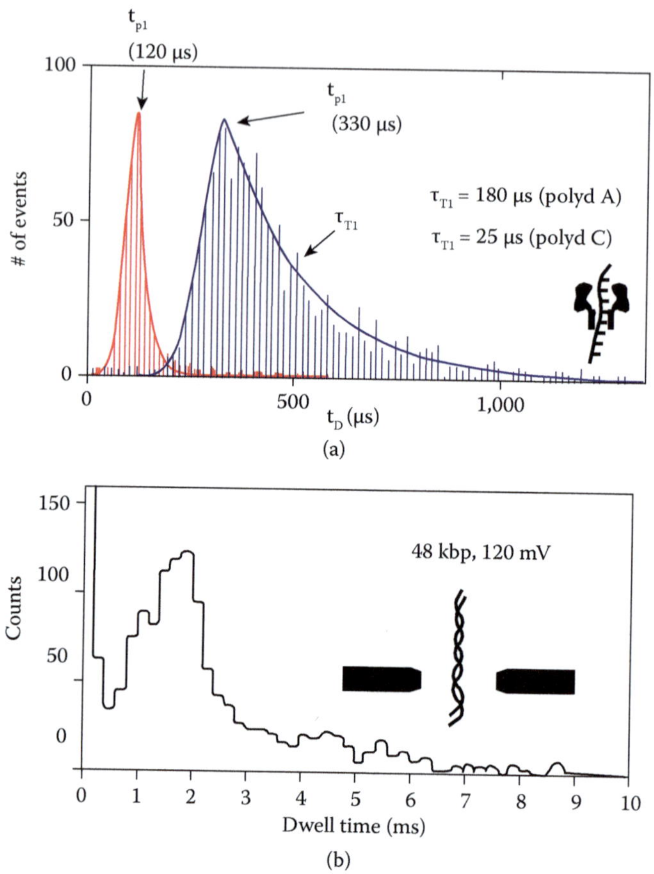

Figure 6.12 DNA transport kinetics. (a) Dwell time distributions for translocations of poly(dA) and poly(dC) ssDNA homopolymers through α-hemolysin protein nanopore, which exhibit different mean timescales (t_P) and tail shapes due to different interactions with the protein pore. (From Meller, A., et al., *Proc. Natl. Acad. Sci. U S A*, 97, 1079–1084, 2000. With permission.) (b) Dwell time distributions for linear lambda-DNA (double-stranded) translocating through a 10 nm SiO_2 pore reveal a very fast population in the first bin at <0.1 ms and a tailing distribution that peaks at 2 ms. (From Storm, A. J., et al., *Phys. Rev. E*, 71, 051903, 2005. With permission.)

profiles of DNA molecules. Under the conditions of these experiments, which were carried out in high LiCl buffer, researchers found that there was a 5% reduction of the velocity of DNA as it transverses the pore (see Figure 6.13d). However, while this reduction seems small in magnitude, a look at the spread in transit times from molecule to molecule reveals the unavoidable diversity in transport times, which is a direct result of the complexity of polymer translocation.

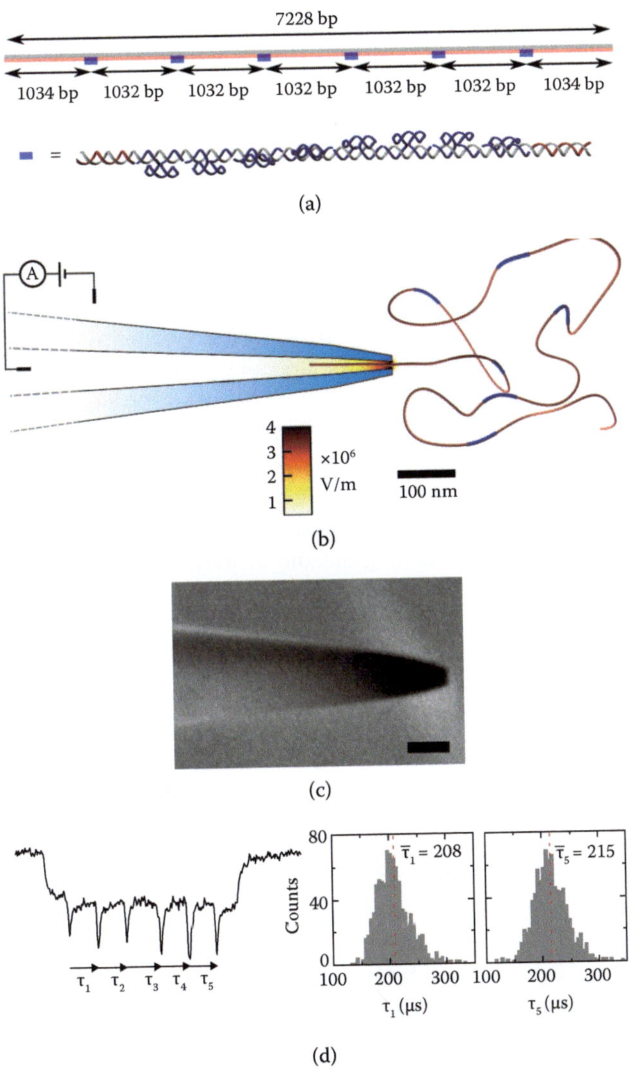

Figure 6.13 DNA transport velocity in large pores. (a) Linear DNA origami construct for testing DNA velocity through nanopipette. (b) Experimental scheme, showing to-scale electric field simulation at tip of pipette. (c) SEM image of pipette. (d) Definitions of time intervals between contrast markers (left), and distribution of early translocation and late translocation times (slower), showing that a DNA molecule slows down as it goes through a pore. (From Bell, N. A., and Keyser, U. F., *ArXiv e-prints* [Online], 1607. http://adsabs.harvard.edu/abs/2016arXiv160704612B, 2016.)

It is important to consider polymer conformation, flexibility, and relaxation, when thinking of the translocation process. The Zimm relaxation time of a polymer, given by

$$\tau_{Zimm} = \frac{0.3\eta\left(\sqrt{N}l_0\right)^3}{K_B T}$$

provides an idea of the timescale of equilibration of the polymer from a perturbed state. When DNA is being sucked into the pore by the electric field, it is perturbed at every stage of the translocation process. If relaxation is faster than transport, then one would expect the transport properties to be dependent on the initial polymer coil dynamics. Under the conditions of most nanopore experiments (1M KCl, pH 7–8), DNA Kuhn length is 90–100 nm, in which case the relaxation time of a DNA coil that is 10 μm long is 70 ms, whereas DNA translocations at these lengths are typically more than an order of magnitude faster, or about 5 ms. Without equilibration, kinetics of the translocating DNA molecule are highly dependent on the initial conformation of the molecule before it entered the pore, which explains molecule-to-molecule diversity and the overall spread in transport times.

Nanopore diameter is also an important consideration that affects DNA transport dynamics. Wanunu and coworkers recently found that decreasing the electric field magnitude and the nanopore diameter to below 3 nm resulted in better-regulated transport than experiments in which the nanopore was merely 1 nm larger (Carson et al. 2015). The underlying cause for this observation is illustrated in Figure 6.14a. Since the electric field near the pore mouth is significant even for a DNA-threaded pore, the DNA coil, which is in close proximity to the pore during a threading event, can be electrokinetically drawn to the pore mouth. This effect leads to autointerference, in which the same DNA molecule that is translocating across the pore interferes with its own transport. The consequence of this phenomenon is that the distribution of transport times (or dwell times) is broadened by long-life events that have a much longer characteristic timescale than they should be, based on electrophoresis alone (see red bars in the distribution on the left). For the smaller pore, in contrast, most of the electric field is extinguished by the threading DNA, because the vacant pore's cross-sectional area is similar to that of the DNA, which leads to most of the ion current being blocked upon DNA entry. In this case, the transport process is more regular, or smooth, as indicated by well-behaved transport time distributions that do not contain many long-lived outlier points. Having this transport characteristic property allows simultaneous detection of a mixture of two DNA lengths. A current versus time trace for a 1:1 mixture is shown in Figure 6.14b. Even the naked eye can distinguish two types of events in this trace, long events (marked by green asterisks) and short events (marked by red asterisks). A dwell time distribution histogram for the mixture is displayed in Figure 6.14c. As seen from the distribution, the histogram contains nearly baseline-separated distributions, one corresponding to the 100 bp and another to the 500 bp DNA molecules. Proof for this was obtained by sampling the two separate molecules using

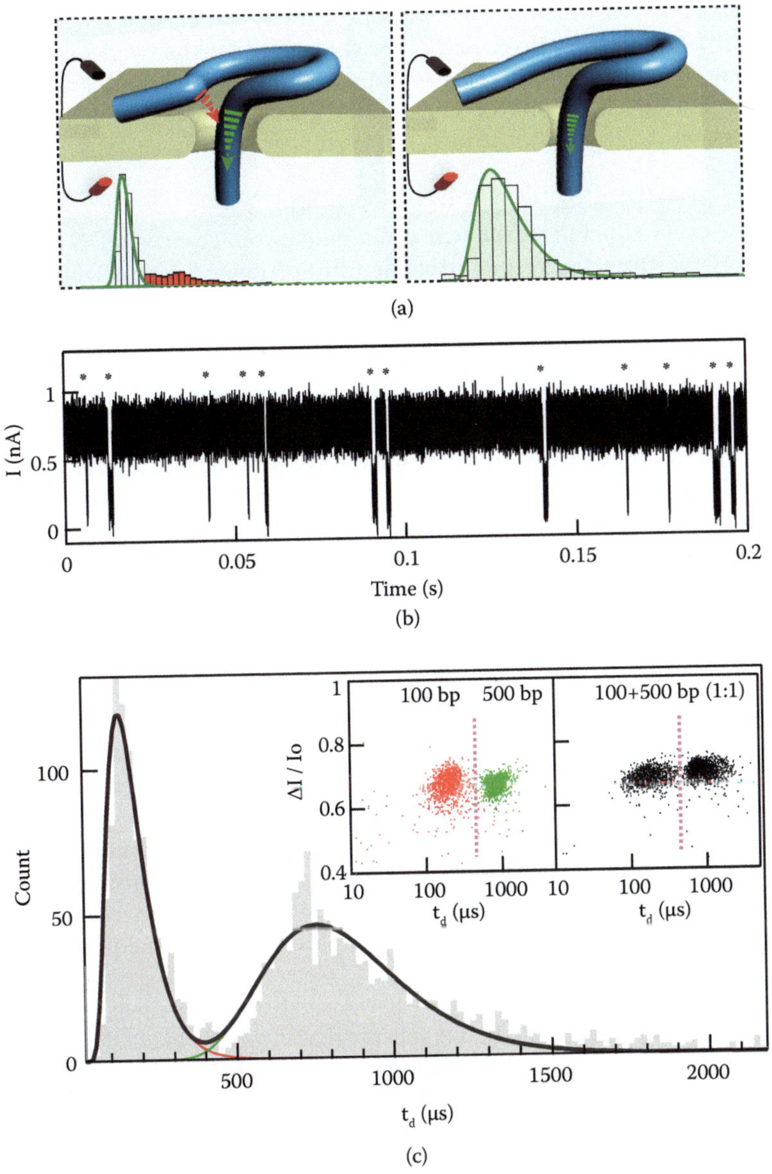

Figure 6.14 Smooth transport in small pores. (a) dsDNA electrophoretically driven through a 4 nm pore (left) and a 3 nm pore (right), as well as experimental distributions of transport times for the same DNA. The distribution for the larger pore has long-lived events (red) due to self-entanglement of the DNA at the pore mouth. (b) Ion current trace of a mixture of 100 bp and 500 bp DNA, in which case longer events correspond to 500 bp molecule. (c) Dwell time distribution for the mixture in (b), showing full resolution of the two lengths. Insets: Scatterplots for control fragments (left) and the mixture (right).

a similar pore, in which the scatterplots of fractional current blockades versus dwell times (shown in inset, left) are well resolved for the two lengths. In comparison, for the 1:1 mixture, the two distributions appear as two well-resolved clouds that match the values obtained for the separate experiments (shown in inset, right).

6.6.3 Enzyme-Regulated DNA Ratcheting

Despite every effort to slow down and regulate voltage-driven DNA transport, one can see from the results discussed in Section 6.6.2 that DNA transport kinetics has many sources of noise associated with random polymer fluctuations and interactions. Although most of the discussion was focused on dsDNA, the situation for a random sequence of single-stranded DNA (ssDNA) is even worse, since this molecule can interact with itself to form transient or stable hydrogen bonds, and thus transport can be stalled to even greater extents than for double-stranded DNA. The case for voltage-driven transport as a method to read DNA molecules is, frankly, hopeless. Fortunately, Mark Akeson and colleagues at University of California–Santa Cruz, invented a method to control DNA motion through a pore base by base (Cherf et al. 2012). In this method, a molecular machine in the form of a DNA-processing enzyme is used to control DNA motion as it is threaded and pulled by voltage applied at a nanopore. This approach solves many issues that are directly relevant for DNA sequencing using nanopores. First, the enzyme serves as a means of stalling a particular base in a pore for a given time, before it moves the DNA to allow readout of the next base. This arrangement allows readout of the current level for a sufficient time for averaging, which improves the accuracy of readout. Second, an enzyme that unwinds dsDNA base by base allows the pore to read only one of the strands (the pore must admit only one strand, not two simultaneously). The distinct advantage of this approach is that a DNA sample can be in a dsDNA state, rather than an ssDNA state, which is much more vulnerable to degrading in solution over time and forming complex structures that preclude it from threading into the pore.

The concept of enzyme-regulated DNA ratcheting is shown in Figure 6.15 (Carson and Wanunu 2015). In one scheme that involves phi29 DNA polymerase (phi29-pol) as the ratchet, DNA that is bound to a phi29-pol is first electrophoretically threaded into the pore, which results in unzipping of a blocking oligomer that was designed to prevent DNA synthesis in solution. Once unzipped, the polymerase can then back-track the DNA through the pore, base by base, by replicating the template strand. (This requires nucleotide triphosphates and divalent ions, typically Mg^{2+}, in the buffer solution.) As each base is incorporated into the growing strand, the template DNA moves up the pore (see Figure 6.15a). Interestingly, during the first step in which electrophoretic force is used to unzip the blocking oligo, a discrete set of signals can be observed that correspond to the sequence of the DNA. This is confirmed by comparing the signal during the force unwinding step to the signal during polymerase-based synthesis, which upon inspection appear as exact mirror images of each other (compare "voltage-driven zipper" to "replication-driven ratchet" in Figure 6.15b). When Jens Gundlach

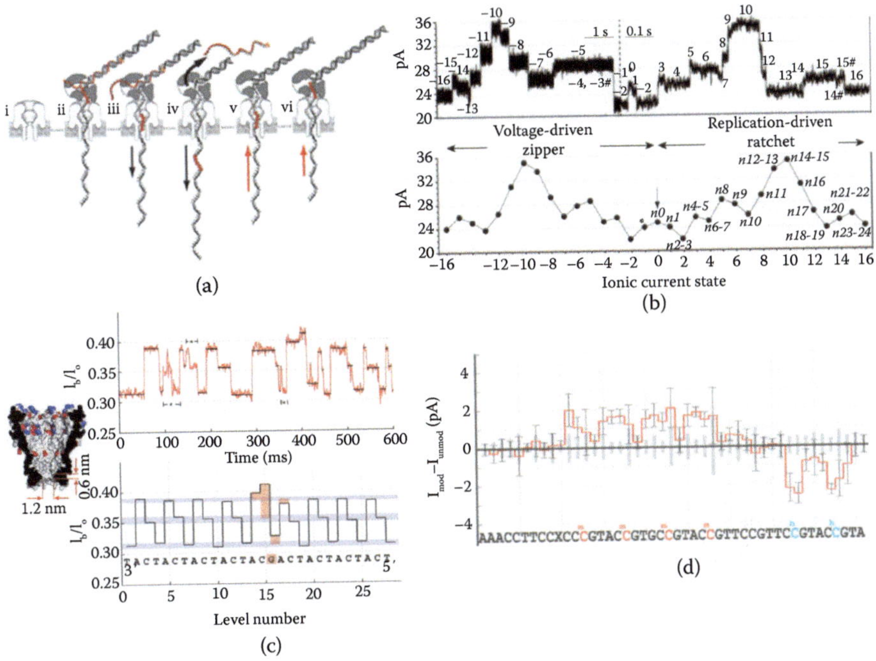

Figure 6.15 (a) Step-by-step sequencing by phi29-pol synthesis of a DNA sequence containing an abasic region: (i) α-hemolysin pore embedded in a lipid bilayer before DNA capture. (ii) Capture of a DNA-polymerase complex by the protruding electric field. (iii) Electrophoretic force starts the unzipping of the blocking oligomer, which prevents DNA digestion and synthesis in the bulk solution. (iv) Once completely unzipped, the blocking oligo diffuses away from the pore. (v) DNA synthesis by phi29-pol reverses the direction of DNA motion against the applied bias. (vi) Once the abasic region of the DNA template reaches the polymerase, synthesis concludes and the DNA-polymerase complex dissociates from the pore mouth. (b) *Top*: Sample current trace of electrophoretic DNA unzipping followed by synthesis by phi29-pol. *Bottom:* When averaging multiple current traces, multiple current levels are discerned that represent an averaged combination of the bases lying within the pore constriction (Cherf et al. 2012). (c) Using phi29-pol to ratchet a repeating CAT DNA sequence with a single substitution of G through a modified MspA nanopore. Left: The narrower and shorter constriction of MspA yield greater spatial resolution for base recognition. Right: Despite the inherent stochastic motion of the DNA sequence, the raw and analyzed current traces detect an increase in fractional current blockade at the location of the guanine substitution (Manrao et al. 2012). (d) By driving DNA through a modified MspA nanopore by polymerase catalysis, the epigenetic modifications methylcytosine (mC) and hydroxymethylcytosine (hmC) can be distinguished from cytosine (C), despite not having single base resolution (Laszlo et al. 2013). (From Carson, S., and Wanunu, M., *Nanotechnology*, 26, 074004–074004, 2015, without modification.)

and his team later repeated the Akeson experiment using a superior nanopore to hemolysin, MspA, the current signals from the DNA sequence appear much more well resolved (Manrao et al. 2012). This is due to the thinner geometry of the nanopore, which allows superior readout (see Figure 6.15c). An immediate application of this method for epigenetic analysis is shown in Figure 6.15d, in which control DNA samples that contain the modified bases 5-methylcytosine (mC) and 5-hydroxymethylcytosine (hmC) were distinguished (Laszlo et al. 2013). Undoubtedly, the enzyme-driven motion of DNA is the method of choice for sequencing, as it allows high-precision measurements of each DNA sequence in the pore at a given time. Other enzymes can be used to achieve this regulated motion of DNA. DNA helicases used for base-by-base unwinding, such as Hel308 (Derrington et al. 2015), for example, have demonstrated comparable or better results to the polymerase method.

6.7 Outlook

The volume of work and the popularity of nanopores have exploded in recent years, underscoring the utility of these objects for many areas of chemistry, physics, and materials science. Nanopores allow researchers to explore transport physics in conditions that are drastically different from the usual conditions in bulk aqueous solutions. Nanopores, in particular, carbon nanotube pores, have allowed researchers to probe effects of confinement, molecular ordering, and surface charge on the efficiency and selectivity of transport. At the same time, nanopore research has contributed to the development of rather practical solutions, with the main example being the spectacular progress achieved in developing nanopore-based DNA sequencing. This process illustrates how the unique physical features of the transport process could be coupled with materials and protein engineering to build a process that achieves a remarkable analytical feat of reading compositional sequence from a single molecule. As we go forward, nanopore technology will continue to develop and thrive. As our abilities to engineer materials at the nanoscale increase, we can anticipate that nanopore technology would be incorporated into large-scale membrane systems that would take advantage of the nanopore transport efficiency and selectivity. Nanopore-based sequencing will continue to develop, and several key challenges to look forward to can be addressed: first, it remains to be seen how many chemical modifications to DNA can be detected, and with what accuracy. Modified cytosines, as well as adenines, are now recognized as significantly important epigenetic marks. Standard sequencing technologies cannot distinguish these chemical modifications in a sequence context, instead relying on bulk analysis of digested DNA and spectrometry/electrophoresis. Second, extending DNA read lengths to the megabase regime would be of immense importance, particularly for sequencing difficult regions that contain repeats and large genomic variations. Third, increasing the sensitivity of nanopore-based sequencing to the single-cell levels would make a great tool for studying genomics at the microscale, and nanopores certainly have the potential to achieve this, given their single molecule detection ability. In addition to DNA sequencing, direct RNA sequencing would provide great insight into the transcriptome and other RNA molecules in the cell to be interrogated. Finally, another major frontier for this technology could be achieving direct readout of protein sequence, which would revolutionize proteomics and allow an unprecedented view into the machinery that makes life tick. This last challenge is especially difficult, as there are many more building blocks that make up proteins (>20 amino acids and their modified versions). Undoubtedly, nanopores are still as relevant and promising as they have been in recent years, and they surely will continue to influence a multitude of areas ranging from transport physics to medical diagnostics.

References

Adam, S. A. 2001. The nuclear pore complex. *Genome Biology,* 2, 1.

Agre, P., Bonhivers, M. & Borgnia, M. J. 1998. The aquaporins, blueprints for cellular plumbing systems. *Journal of Biological Chemistry,* 273, 14659–14662.

Agre, P., Borgnia, M. J., Yasui, M., Neely, J. D., Carbrey, J., Kozono, D., Beitz, E., Hoffert, J., Leitch, V. & King, L. S. 2001. Discovery of the aquaporins and their impact on basic and clinical physiology. *Curr. Topics. Membr.* 51, 1–38.

Aidley, D. J. & Stanfield, P. R. 1996. *Ion Channels: Molecules in Action*, New York Cambridge University Press.

Alberts, B., Bray, D., Lewis, J., Raff, M., Roberts, K. & Watson, J. D. 2007. *Molecular Biology of the Cell,* New York, Garland Science.

Barboiu, M. 2012. Artificial water channels. *Angewandte Chemie International Edition,* 51, 11674–11676.

Bell, N. A. W. & Keyser, U. F. 2016. Direct measurements reveal non-Markovian fluctuations of DNA threading through a solid-state nanopore. *ArXiv e-prints [Online], 1607.* Available: http://adsabs.harvard.edu/abs/2016arXiv160704612B [Accessed July 1, 2016].

Bocquet, L. & Barrat, J.-L. 2007. Flow boundary conditions from nano-to microscales. *Soft Matter,* 3, 685–693.

Bocquet, L. & Charlaix, E. 2010. Nanofluidics, from bulk to interfaces. *Chemical Society Reviews,* 39, 1073–1095.

Brewer, M. L., Schmitt, U. W. & Voth, G. A. 2001. The formation and dynamics of proton wires in channel environments. *Biophysical Journal,* 80, 1691–1702.

Buyukdagli, S., Manghi, M. & Palmeri, J. 2010. Ionic capillary evaporation in weakly charged nanopores. *Physical Review Letters,* 105, 158103.

Buyukdagli, S., Manghi, M. & Palmeri, J. 2011. Ionic exclusion phase transition in neutral and weakly charged cylindrical nanopores. *The Journal of Chemical Physics,* 134, 074706.

Carson, S. & Wanunu, M. 2015. Challenges in DNA motion control and sequence readout using nanopore devices. *Nanotechnology,* 26, 074004–074004.

Carson, S., Wilson, J., Aksimentiev, A. & Wanunu, M. 2015. Smooth DNA transport through a narrowed pore geometry. *Biophysical Journal,* 107, 2381–2393.

Cherf, G. M., Lieberman, K. R., Rashid, H., Lam, C. E., Karplus, K. & Akeson, M. 2012. Automated forward and reverse ratcheting of DNA in a nanopore at five angstrom precision(). *Nature Biotechnology,* 30, 344–348.

Cottin-Bizonne, C., Jurine, S., Baudry, J., Crassous, J., Restagno, F. & Charlaix, É. 2002. Nanorheology: An investigation of the boundary condition at hydrophobic and hydrophilic interfaces. *The European Physical Journal E-Soft Matter,* 9, 47–53.

Craig, V. S. J., Neto, C. & Williams, D. R. M. 2001. Shear-Dependent boundary slip in an aqueous newtonian liquid. *Physical Review Letters,* 87, 54504.

Dai, H. J. 2002. Carbon nanotubes: Synthesis, integration, and properties. *Accounts of Chemical Research,* 35, 1035–1044.

De Groot, B. L. & Grubmüller, H. 2005. The dynamics and energetics of water permeation and proton exclusion in aquaporins. *Current Opinion in Structural Biology,* 15, 176–183.

Dellago, C., Naor, M. M. & Hummer, G. 2003. Proton transport through water-Filled carbon nanotubes. *Physical Review Letters,* 90, 105902.

Derrington, I. M., Craig, J. M., Stava, E., Laszlo, A. H., Ross, B. C., Brinkerhoff, H., Nova, I. C., et al. 2015. Sub-angstrom single-molecule measurements of motor proteins using a nanopore. *Nature Biotechnology,* 33, 1073–1075.

Dresselhaus, M. S. & Saito, R. 1998. *Physical Properties of Carbon Nanotubes,* London, Imperial College Press.

Faller, M., Niederweis, M. & Schulz, G. E. 2004. The structure of a mycobacterial outer-Membrane channel. *Science,* 303, 1189–1192.

Fyles, T. 2007. Synthetic ion channels in bilayer membranes. *Chemical Society Reviews,* 36, 335–347.

Geng, J., Kim, K., Zhang, J., Tunuguntla, R., Comolli, L., Allen, F., Cho, K., et al. 2014. Stochastic transport through carbon nanotubes in lipid bilayers and live cell membranes. *Nature,* 514, 612–615.

Gravelle, S. 2016. *Nanofluidics: A pedagogical introduction.* Available: http://simon-gravelle.fr/documents/NanofluidicsPedagogicalIntroduction.pdf [Accessed March 10, 2016].

Hanasaki, I. & Nakatani, A. 2006. Flow structure of water in carbon nanotubes: Poiseuille type or plug-like? *The Journal of Chemical Physics,* 124, 144708.

Holt, J. K., Park, H. G., Wang, Y., Stadermann, M., Artyukhin, A. B., Grigoropoulos, C. P., Noy, A. & Bakajin, O. 2006. Fast mass transport through sub-2-Nanometer carbon nanotubes. *Science,* 312, 1034–1037.

Hummer, G. 2007. Water, proton, and ion transport: From nanotubes to proteins. *Molecular Physics,* 105, 201–207.

Hummer, G., Rasaiah, J. C. & Noworyta, J. P. 2001. Water conduction through the hydrophobic channel of a carbon nanotube. *Nature,* 414, 188–190.

Ilan, B., Tajkhorshid, E., Schulten, K. & Voth, G. A. 2004. The mechanism of proton exclusion in aquaporin channels. *PROTEINS: Structure, Function, and Bioinformatics,* 55, 223–228.

Jing, B., Janout, V., Herold, B., Klotman, M., Heald, T. & Regen, S. 2004. Persulfated molecular umbrellas as anti-HIV and anti-HSV agents. *Journal of the American Chemical Society,* 126, 15930.

Kalra, A., Garde, S. & Hummer, G. 2003. Osmotic water transport through carbon nanotube membranes. *Proceedings of the National Academy of Sciences,* 100, 10175–10180.

Kasianowicz, J. J., Brandin, E., Branton, D. & Deamer, D. W. 1996. Characterization of individual polynucleotide molecules using a membrane channel. *Proceedings of the National Academy of Sciences,* 93, 13770–13773.

Kim, K., Geng, J., Tunuguntla, R., Comolli, L. R., Grigoropoulos, C. P., Ajo-Franklin, C. M. & Noy, A. 2014. Osmotically-Driven transport in carbon nanotube porins. *Nano Letters,* 14, 7051–7056.

Kotsalis, E. M., Walther, J. H. & Koumoutsakos, P. 2004. Multiphase water flow inside carbon nanotubes. *International Journal of Multiphase Flow,* 39, 995–1010.

Kumar, M., Grzelakowski, M., Zilles, J., Clark, M. & Meier, W. 2007. Highly permeable polymeric membranes based on the incorporation of the functional water channel protein Aquaporin Z. *Proceedings of the National Academy of Sciences,* 104, 20719–20724.

Kwok, H., Briggs, K. & Tabard-Cossa, V. 2014. Nanopore fabrication by controlled dielectric breakdown. *PLoS One,* 9, e92880.

Laszlo, A. H., Derrington, I. M., Brinkerhoff, H., Langford, K. W., Nova, I. C., Samson, J. M., Bartlett, J. J., Pavlenok, M., & Gundlach, J. H. 2013. Detection and mapping of 5-methylcytosine and 5-hydroxymethylcytosine with nanopore MspA. *Proceedings of the National Academy of Sciences,* 110, 18904–18909.

Majumder, M., Chopra, N., Andrews, R. & Hinds, B. J. 2005. Nanoscale hydrodynamics: Enhanced flow in carbon nanotubes. *Nature,* 438, 44–44.

Manrao, E. A., Derrington, I. M., Laszlo, A. H., Langford, K. W., Hopper, M. K., Gillgren, N., Pavlennok, M., Niederweis, M. & Gundlach, J. H. 2012. Reading DNA at single-Nucleotide resolution with a mutant MspA nanopore and phi29 DNA polymerase. *Nature Biotechnology,* 30, 349–353.

Mathé, J., Aksimentiev, A., Nelson, D. R., Schulten, K. & Meller, A. 2005. Orientation discrimination of single-Stranded DNA inside the α-hemolysin membrane channel. *Proceedings of the National Academy of Sciences of the United States of America,* 102, 12377–12382.

Matile, S., Som, A. & Sorde, N. 2004. Recent synthetic ion channels and pores. *Tetrahedron,* 60, 6405–6435.

Meller, A., Nivon, L., Brandin, E., Golovchenko, J. & Branton, D. 2000. Rapid nanopore discrimination between single polynucleotide molecules. *Proceedings of the National Academy of Sciences of the United States of America,* 97, 1079–1084.

Murata, K., Mitsuoka, K., Hirai, T., Walz, T., Agre, P., Heymann, J. B., Engel, A. & Fujiyoshi, Y. 2000. Structural determinants of water permeation through aquaporin-1. *Nature,* 407, 599–605.

Murray, C., Shabany, H. & Gokel, G. 2000. The central relay unit in hydraphile channels as a model for the water-and-ion 'capsule' of channel proteins. *Chemical Communications,* 2000, 2371–2372.

Noy, A., Park, H.-G., Fornasiero, F., Holt, J., Grigoropulous, C. & Bakajin, O. 2007. Nanofluidics in carbon nanotubes. *Nano Today,* 2, 22–29.

Plesa, C., Van Loo, N., Ketterer, P., Dietz, H. & Dekker, C. 2015. Velocity of DNA during translocation through a solid-State nanopore. *Nano Letters,* 15, 732–737.

Pomes, R. & Roux, B. 1996. Structure and dynamics of a proton wire: A theoretical study of H+ translocation along the single-file water chain in the gramicidin A channel. *Biophysical Journal,* 71, 19–39.

Schlichting, I., Berendzen, J., Chu, K., Stock, A. M., Maves, S. A., Benson, D. E., Sweet, R. M., Ringe, D., Petsko, G. A. & Sligar, S. G. 2000. The catalytic pathway of cytochrome P450cam at atomic resolution. *Science,* 287, 1615–1622.

Schoch, R. B., Han, J. & Renaud, P. 2008. Transport phenomena in nanofluidics. *Reviews of Modern Physics,* 80, 839.

Singer, A., Rapireddy, S., Ly, D. H. & Meller, A. 2012. Electronic barcoding of a viral gene at the single-Molecule level. *Nano Letters,* 12, 1722–1728.

Song, L., Hobaugh, M. R., Shustak, C., Cheley, S., Bayley, H. & Gouaux, J. E. 1996. Structure of staphylococcal α-hemolysin, a heptameric transmembrane pore. *Science,* 274, 1859–1865.

Storm, A. J., Chen, J. H., Zandbergen, H. W. & Dekker, C. 2005. Translocation of double-Strand DNA through a silicon oxide nanopore. *Physical Review E,* 71, 051903.

Striolo, A. 2006. The mechanism of water diffusion in narrow carbon nanotubes. *Nanoletters,* 6, 633–639.

Sui, H., Han, B.-G., Lee, J. K., Walian, P. & Jap, B. K. 2001. Structural basis of water-Specific transport through the AQP1 water channel. *Nature,* 414, 872–878.

Tunuguntla, R., Allen, F., Kim, K., Bellivieau, A. & Noy, A. 2016. Ultrafast proton transport in sub-1-nm diameter carbon nanotube porins. *Nature Nanotechnology,* 11, 639–644.

Tunuguntla, R., Henley, R. Y., Yao, Y.-C., Pham, T.A., Wanunu, M., Noy, A. 2017. Enhanced water permeability and tunable ion selectivity in sub-nanometer carbon nanotube porins. *Science,* 357, 792–796.

Wanunu, M. 2012. Nanopores: A journey towards DNA sequencing. *Physics of Life Reviews,* 9, 125–158.

Index

A

B

I

L

M

N

O

P

Q

R

S

T

U

V

W

X

Z